On the Banks of the Gaṅgā

On the Banks of the

GAṄGĀ

When Wastewater Meets a Sacred River

Kelly D. Alley

THE UNIVERSITY OF MICHIGAN PRESS
Ann Arbor

Published in the United States of America by
The University of Michigan Press
Manufactured in the United States of America
♾ Printed on acid-free paper

2010 2009 2008 2007 5 4 3 2

A CIP catalog record for this book is available from the British Library.

Library of Congress Cataloging-in-Publication Data

Alley, Kelly D., 1961–
On the banks of the Ganga : when wastewater meets a sacred river / Kelly D. Alley.
p. cm.
Includes bibliographical references and index.
ISBN 0-472-09808-X (cloth : alk. paper) — ISBN 0-472-06808-3 (pbk. : alk. paper)
1. Nature—Effect of human beings on—Ganges River Valley (India and Bangladesh) 2. Water—Religious aspects—Hinduism. 3. Water—Pollution—Ganges River Valley (India and Bangladesh) 4. Sewage—Ganges River Valley (India and Bangladesh) 5. Water quality—Ganges River Valley (India and Bangladesh) 6. Vàrànasi (Uttar Pradesh, India)—Religious life and customs. 7. Vàrànasi (Uttar Pradesh, India)—Environmental conditions. 8. Ganges River Valley (India and Bangladesh)—Religious life and customs. 9. Ganges River Valley (India and Bangladesh)—Environmental conditions I. Title.

GF662.G36 A45 2002
304.2'8'0954—dc21 2002075015

ISBN 978-0-472-09808-8 (cloth : alk. paper)
ISBN 978-0-472-06808-1 (pbk. : alk. paper)

For Ali, Zahra, and Khayr

Contents

Illustrations

Maps

Figures

Acknowledgments

AS I TURN TO THE FINAL TASK of this book project, to writing an acknowledgment that gives thanks to all the wonderful people who have helped me along the way, I feel a bit self-critical. Alas, has this turned out to be just another discourse on Gaṅgā, that only speaks to the problems but fails to solve them? I know that for all the support from my Indian colleagues over the years, this will be their overriding question. Indeed it is a good one. So I write this section, not with the intent to celebrate my own achievement in ethnographic writing but to highlight those who have been concerned for years with these problems and who have helped me to see their many dimensions. These people are the guides and inspiration for this book.

I remember a conversation I had with a British activist several years ago in which I explained my reason for sustaining this focus over the long term. The real reason, I told him, is that I like and admire the people I have been writing about—especially the activists who struggle in various ways against insurmountable odds and the ritual specialists whose live connections to Gaṅgā I will never fully comprehend. My interest to return, year after year, to this often depressing story line about wastewater and the Ganga Action Plan has been sustained by their very incredible stories and life experiences. I have lived my own activist urges vicariously through their much more exceptional practices.

Those who have helped me most are all mentioned in this book: Rakesh Jaiswal, M. C. Mehta, Veer Bhadra Misra (Mahantji), S. N. Upadhyay, and S. K. Misra, among many others. Bailey Green, P. K.

Jha, and Michael Duffy also provided valuable information on grass-roots organizing, water quality monitoring and wastewater management technologies. All these activists and scientists were open with me about their perspectives and opinions, making a study of discourse very easy and practical. They also took an interest in evaluating this exercise and providing their critiques of my work in friendly, helpful ways.

I also appreciate the assistance provided by residents of Daśāśvamedha, by Anil Yadav, Raju Kumar and Prema Tiwari, S. K. Misra (Munnanji), Kishore Raman Dube (Babu Maharaj), and Lal Baba. Shobnath Yadav undertook with me the first major survey of Daśāśvamedha ghāṭ and became very involved in fleshing out the first stage of this research. All those we surveyed on the ghāṭ tolerated our curiosities and intrusions and gave time and energy to make them into sincere forms of "transcultural exchange." My deepest affections go to the people of Daśāśvamedha.

For my scholastic excursions, I am thankful to the direction provided by Owen Lynch, who actually read all my publications and knew exactly what I needed to do to improve the project. I have also been helped along the way by Anne Feldhaus, Lance Nelson, Ginny Van Dyke, Subhadra Channa, Rana P. B. Singh, Susanne Heober Rudolph, Lloyd Rudolph, Ann Gold, and David Haberman. Their valuable insights into Indian politics and religion gave direction to my research concerns. Outside of the community of South Asianists, many colleagues assisted in making my particular project part of a wider network of concerns, and helped me to feel professionally linked to that wider net. Emiko Ohnuki-Tierney, J. Peter Brosius, Mark Chernaik, Michele Masucci, Paul Starr, and Richard Perritt connected me to geography, sociology, law, and symbolic and environmental anthropology. I also thank the former Chair of my department, Gregory Kowalski, and our office administrator, Becky Gregory, for enduring an everyday understanding of this project and keeping an open mind about the nomadic practices of anthropologists. Jay Winks helped turn a theoretical idea into useful charts. Mary Cameron and Jack Schmidt also helped edit sections of this book and presented me with critical prods. Finally, I am grateful to Conner Bailey and Charles Faupel, colleagues at Auburn who introduced me to Alabama's waste politics and gave me an early orientation to this social science focus. I also appreciate the insightful comments provided by the anonymous reviewers of the University of Michigan Press.

The fieldwork could not have been accomplished without the generous fellowships provided by the American Institute of Indian Studies, Auburn University's Research grant-in-aid and Humanities grant programs, and the U.S. Information Agency (now U.S. Department of State), Office of Citizen Exchanges. The Citizen Exchanges grant, while focused on the facilitation of exchanges between lawyers, environmental activists, and scientists, gave me the opportunity to complete the final section of this book.

Not least of all, I am grateful to my family, and especially to my husband, daughter, and son, for enduring many hikes to the riverbank and long tours of *nālās* (wastewater drains) and treatment plants. "Not another nālā!" my daughter would say. Near and away from nālās, my husband kept my feet on the ground, reminding me of the very critical view of everyday life that only a resident and citizen can have. Despite the fact that I wove in and out of this view and enjoyed temporary visits to the country and field sites, my husband and many informants accepted that mobility as a part of my work. Without their support and tolerance, I would have learned very little.

Kelly D. Alley
Auburn, Alabama

Introduction: Field Methods and Layers of Data

I STARTED THIS PROJECT about nine years ago, while studying tourism and public culture in north India. I remember the day that the original idea for this study emerged from a dialogue I overheard while touring the north Indian city of Banaras with several American and German tourists. While boating along the river Gaṅgā and watching Hindu pilgrims bathe in this sacred river, these tourists discussed the state of the river, expressing horror that the native population did not recognize its polluted nature. One said, "I wouldn't put my big toe into this river, it is so polluted!" As an observer, I began to think about this Western tourist interpretation in light of the Hindu practice of bathing in sacred rivers, and about how this view of pollution is juxtaposed against a view of the sacred. Other questions emerged. Are these differences in thinking and approach mirrored within and outside of India in other ways? How should an investigation of these two perspectives, the one in relation to the other, proceed? What shall the methodology be?

Since the very beginning of the project, I have been guided by very few prior assumptions about how to build a road map for integrating diverse fields of data and human experience. To respond to the comments tourists were making, I began a process of data collection that, in hindsight, appears to have grown by way of an outward layering of data sets. I began this journey at the very place where this conversation began, at a sacred space bordering this reach of the Gaṅgā. I chose to locate my first study of native interpretations of the river Gaṅgā at the site of Daśāśvamedha, an important sacred space in the city of Banaras (known officially as Varanasi). During the summer months of 1993 and 1994, I elicited responses to a set of questions from those who worked in pilgrimage practices and trades on the sacred space that adjoins this river, known in Hindi as a *ghāṭ*. I designed this questionnaire with members of a local nongovernmental organization (NGO) and with residents of Daśāśvamedha. A research assistant and I cataloged the audiotaped and written responses, most of which were communicated in Hindi, and began to appreciate these as texts of dense meaning. In a cumulative fashion, they formed the foundation for my journeys outward, conceptually, culturally, and spatially, to other cities in the Gangetic Plain.

While carrying out the survey at Daśāśvamedha, I began to understand how the political hierarchy of life on the ghāṭs and a place-based notion of power affected residents' relations with the river. The family histories were fascinating, and interviews with elders, ritual specialists, service-providers, women of local households, and transient pilgrims were engaging. In all these, I guided my curiosity with a set of questions that explored native understandings of dirtiness, sacred purity and impurity, and environmental pollution. In other questions, I asked respondents to define their work space and their sacred space, list specific elements or instances of waste, and describe their understanding of the city's waste management system.

While carrying out these open-ended interviews, I went about trying to trace the city's wastewater flows, by walking along the river bank and visiting points of effluent disposal and collection. Initially, the best guides for this investigation proved to be the Water and Engineering Board's maps of the city sewage system and a few colonial documents on the construction of the first drainage system for this urban area. With these grids as guides, I went about mapping major and minor flows, noting their input and output points, and collecting studies of the physical and chemical nature of wastewater at multiple junctures. One morning toward the end of my first summer of research, I was taken on a particularly interesting tour by a former employee of the Water and Engineering Board (Jal Nigam) of Banaras. By motorcycle, this former employee took me to all the bypass drains for various government and industrial effluent treatment plants. A bypass drain, which diverts wastewater from a treatment plant or pumping station directly into the Gaṅgā, is used to channel wastewater toward the river when the pumping or treatment stations are not working or when there is a shortage of power to run the units. Bypass drains opened my eyes to the underbelly of the official system and to the escape valves that officials and industrialists use to give the appearance of functioning treatment units. However, I soon found that these bypasses, though hidden from view on land, were clearly evident to those who ply the river by boat or who sit on the riverbank supervising and conducting ritual practices. I found that this vantage point gave boatmen and pilgrim priests a more critical understanding of the wastewater management system in the city than that held by other Banaras residents.

After two years, I expanded my study beyond the community at Daśāśvamedha and the city of Banaras to grapple in a more direct way with different interpretations and uses of the river. I began to feel the tensions and chasms between pilgrim priests' and boatmen's interpretations of the Gaṅgā, interpretations embedded in environmental policies and legal documents, and those associated with waste management practices. I visited wastewater management and water and sanitation offices at all government levels to get at official visions of the river and to understand technological applications for managing wastewater. I visited local offices of the Uttar Pradesh Jal Nigam (the Uttar Pradesh Water and Engineering Board), the Jal Sansthan (Water Board), municipal bodies, and the Ganga Project Directorate in the Central Ministry of Environment and Forests to discuss the objectives of the government's first environmental program, the Ganga Action Plan. I also collected testimonies and court documents to assess the functioning of the Central and State Pollution Control Boards. As my focus grew outward geographically, I was able to compare pollution discourses and wastewater management projects in other cities in the state of Uttar Pradesh, in Kanpur, Allahabad, Haridwar, and Rishikesh and link them to assessments made in overseas development offices.

At almost every stage, I met and talked with members of NGOs involved with river pollution prevention and river ecology restoration. I also met with religious leaders and *sādhu-sants* (world renouncers-saints) to discuss their views of Gaṅgā and their knowledge of environmental pollution and governmental and nongovernmental waste management projects. I call this an *outward layering methodology* because I began with a problem and then traced the relations between individuals and groups who appear to create the problem, define and debate the problem from multiple perspectives, and manipulate or attempt to solve the problem. In this case, the problem was a paradox for some and a moral predicament for others: How is it that a sacred river can be polluted? Can it be polluted? The outward layering of field sites led to a widening of geographic space and to a more diverse sphere of cultural discourses and practices. This layering also made the integration of data from these various fields an ever-present challenge.

While continuing to look at the cultural worldviews of Daśāśvamedha residents and government officials working in waste management positions, I started in 1995 to look at legal debates about the river. I searched for a hook in the past, some set of orders and directions, that might reveal colonial views of water and sanitation and provide a beginning to this institutional history. At that time, I unearthed fascinating historical documents in the dusty law libraries of Delhi. I soon found that these documents provided a window into colonial discourses on water resources and revealed some of the conceptual antecedents to contemporary institutional structures.[1]

To sustain this outward layering methodology, I used several data collection techniques. First, I became a participant observer during court sessions, meetings, ritual events, festivals, and household functions held by officials, activists, and religious leaders. To observe ecological and urban settlement conditions, I toured reaches of the river and the technological facilities covered in government policies and nongovernmental and legal reports. To understand the qualities and power of sacred space, I visited sacred grounds and talked with ritual specialists and religious leaders. After collecting written documents of many types—petitions, orders, affidavits, agency reports, and policy documents—I was able to interpret them alongside interpretations that emerged from songs and poems, ritual practices and key symbols. I discuss pieces of these data sets throughout the book and use them to analyze conceptions of environmental problems and problem solving and specific understandings of cultural and political contingencies.[2]

Reflexive Caveats

Reflexive caveats are always in order while doing fieldwork. Dilemmas arise, puzzles confound, and social relations push the limits of one's familiar world. I would like to list a few here to give the reader a sense of what the integration of data has meant to the researcher in a more intimate way. How can data be collected and understood across competing fields of influence when the researcher is pulled between different sites and cultural groups?

When moving from the sacred space of Daśāśvamedha to the corridors of government buildings and then into the crowded spaces of the courtroom, the problem of integrating levels was apparent and pressing. In fact, to integrate levels of data I had to engage in a fair amount of crossing-over activities. This meant participating in many distinct cultural spheres, exploring an array of physical and sacred spaces, and assessing the flow and quality of the river at many junctures. I will point out a few crossing-over experiences here to highlight the personal problems I encountered while following the outward layering methodology.

In the early stages of my field research, government officials spent a good deal of their time discussing sensitive problems of sewage management with me, and members of NGOs welcomed me to their meetings. During the first few years in particular, officials of the Uttar Pradesh Jal Nigam provided tours of waste treatment plants in each city and gave me valuable maps of the city sewage system. But I also met environmentalists and citizens who warned that everything looks *pakkā* (neat and accurate) on paper. They urged me to get at more than tours and maps, more than official representations of physical plants and actual projects. But this crossing-over, this movement from the public space of the ghāṭ to the meeting places of NGO members and then into government offices and industrial plants, was only partially achieved, and after a considerable amount of meandering around the periphery. Barred from the everyday activities of bureaucratic decision making and the everyday functions within treatment plants at government or industrial sites, I was forced to view government and industry from their physical borders and from my own vantage points on the banks of the great river Gaṅgā.

With the identity of an ambiguous environmentalist, I began to create a space for my own observational and participatory activities. Despite my marginal status, I found that I could achieve a level of access to government documents that some nongovernmental group members could not. To win the favor of nongovernmental groups, I engaged in subversive forms of data collection by requesting official materials and passing them on to NGOs. While these activities were not particularly radical or even dangerous, I did wonder about how they undermined the image of the neutral anthropologist. As my work progressed, the communication gap between nongovernmental environmental groups and government offices grew wider and made my role as information shuttler slightly more important.

This layering effect also generated obstacles that blocked access to deep data on all sides, including my own.[3] Obstacles to accessing deep data were created when officials and industrialists washed away evidence of their activities in cool crisp rhetoric, or when religious leaders and ritual specialists shifted to their mother tongue and discussed transcendent matters I could not understand. Personal obstacles took the form of ethnocentric impulses that prevented me from fully participating in cultural practices of worship to Gaṅgā, especially *snān* (ritual ablutions).

Postcolonial and postorientalist dilemmas also emerged as I crossed through various perspectives. They were especially evident in discussions I had with Hindu religious leaders, ritual specialists, and pilgrims. While academics warned me not to essentialize the Orient, these citizens wondered about my true beliefs and allegiances and blamed me for remaining spiritually distant. Without regularly performing snān, they pointed out, I was missing the big picture. On one occasion, while I interviewed an important Hindu religious leader and world renouncer, a devotee asked me, "Have you bathed [in Gaṅgā] today?" *(Āpne snān kiyā?)* His message was: how can you, a nonbeliever or *nāstik*, begin the quest for truth without embracing Gaṅgā's sacred purity through snān?

These ethnocentric impulses also led me to question my own disciplinary intentions. I became engaged in a quiet debate about whether Western social science could be or should be used to talk about Indian religion or society. Could the anthropologist's categories achieve the imagined ideal of commensurability that social scientists pose between self and other? Can the anthropological framework ever be commensurate enough with the Hindu, scientific, or bureaucratic framework to do justice to analytic representations? For the religious leaders and ritual specialists I met, this was precisely the problem. Since I sought to understand cultural responses to environmental problems in anthropocentric rather than transcendent terms, I risked missing their whole point. To many, I lacked *bhaktī*, or devotion to the divine, and was not able to achieve an understanding of bliss.[4] Feeling their disapproval, I wondered: If emotion is at play in the relationships that pilgrims, priests, and residents develop with Gaṅgā and I was lacking that, then how should I characterize that essence or force? If motherly love is a dominant metaphor informants use to characterize their bond with Gaṅgā, then how should I feel it?

Nonetheless, I defended myself against the charges that I lacked bhaktī. I agreed that the real experiences of fieldwork had to be explored through immersion in Gaṅgā and through other contacts with her. But admittedly, I found self-realization through negative paths! A bout with viral hepatitis, contracted while drinking tea made from Gaṅgā water, forced me to look at the distinction between external form and devotion in my own mind and contrast it with the distinction understood by my informants. While my bout with hepatitis brought me closer to the fear of "water pollution" and pushed me farther from devotion, it did little to press the connection between environmental pollution and public health in the minds of informants.

When discussing my illness with others, I learned that waterborne diseases did not unequivocally signal the failure of government or stir civic consciousness. Many blamed me rather than the water or sewage infrastructure of the city (which converts Gaṅgā into drinking water and then wastewater) for contracting the illness. I had *piliyā rog* (the yellow sickness) because I had "no immunity," and that, they argued, was my fault. My retorts—Wasn't it a sign of poor government services?—fell on deaf ears. Instead, they laughed, it was rather naive to hold the local municipality ("that useless institution") or central and state agencies ("those corrupt, big fish") accountable for my misfortunes. Wouldn't it be right to hold the local institution accountable and demand that it manage water resources better? Gradually, I learned that remarks about immunity were meant to warn citizens that public offices will not protect them. In times of need and to protect health and well-being, one must turn to the grace of Gaṅgājī, and after that the extended family and religious organizations.[5] Beyond these powers and supports, immunity is the best alternative.

Finally, I learned an important point about social science and knowledge building when I fell victim to the same kinds of representations that make native points of view into commodities for media arguments. When journalists commented on my research and presence at various functions, they tended to misquote me and misrepresent my views. So when I found that informants did not want me to write about them or wanted me to write about them only in certain ways I understood their concerns. I also disliked becoming a signifier in my own story line. Images of the Orient and Occident have their multiple productions and reproductions in scholarship and journalism, and the cultural essences they project have helped to manage and produce the Orient politically, militarily, ideologically, scientifically, and imaginatively for over a century (Said 1978). While the undersides of these discursive productions still have to be explored, a point I will address in the conclusion, the fact remains that the tendency to essentialize is built into the anthropological enterprise, putting noble claims about the ethics of fieldwork on shaky ground. The thought that I might also be vulnerable to such essentialist images helped me to see the anthropological venture from a more critical point of view.

Knowledge building is supposed to lead to information sharing among professionals and the wider public. However, there were times when sharing my writing with informants caused even greater suspicion than hiding it did. Suspicions arose among those who saw my papers and, unable to read them in English, could only imagine their contents. In the current climate of cross-cultural disjunctures, it is practical for others to suspect that they have been misrepresented. The language differences only accentuate the problem. On the other side, the camaraderie that I established on the basis of a shared interest in literacy also had its unstable moments. With environmental activists and lawyers, literacy was valuable to the extent that sharing data and knowledge could generate a certain measure of power in coalition building. But all players knew that individual interests were intersecting to shape the final delivery of the pen.

Ironically, I found solace in an adage a pilgrim priest told me. Rice on a plate and words on a page, he explained, are both the same. If you taste one kernel (or read one word), you have tasted the whole plate (read the whole page). There is little need, he seemed to say, to take what is written "on paper" too literally. One written word is just like another, like one grain of rice to another. There is little chance that words on a page will challenge a network of social relations built along other lines.

Outward Layering Methodology and Multisite Ethnography

The scenarios I would like to bring alive here respond to the following question: What are the public discourses, rules, and practices followed to identify and manage wastewater and its intersections with important rivers? Focusing on the ways Indian residents interact with the river Gaṅgā, a river that stands as the lifeline for inhabitants of the Himalayas and northern plains, I located my field research in four urban centers. But the paths of communication I traced and present here took me beyond these locales to state and national capitals, and to other sites abroad. The four cities I chose are the largest urban centers in the north Indian state of Uttar Pradesh. Residents of these cities express decidedly urban predicaments, though in culturally unique ways.[6]

I explored my questions about the Gaṅgā and city wastewater with residents who maintain a proximity to the river by way of their occupational activities, spiritual pursuits, and livelihood needs. I developed this loose network of informants to mirror what I saw as a cultural configuration. Despite membership in other imagined communities, these individuals appeared to create a level of sociability simply by knowing each other's or *about* each other's involvement with the river.[7] Like the intersecting discourses described in Foucault's *I, Pierre Rivière* (1975), they were voices that did not speak in isolation. Their manner of knowing and speaking was strikingly similar to the manner Brass (1997:59) describes in his analysis of debates on riots in northern India. Participants in these discourses seemed to be aware of the existence of entirely different voices and were able to switch voices themselves. The layering effect I was creating through fieldwork was, in other words, present in citizens' daily lives. So my goal had to include an understanding of the layering effect in citizens' terms as well.

The citizens I introduce belong to a diverse array of professions and hereditary traditions; they are of different castes, religions, and disciplines. On the basis of their professional or hereditary titles, they can be grouped under the following categories: Hindu religious leaders, pilgrim service providers, other ritual specialists, environmental activists, lawyers, scientists, and government officials. Many of these individuals hold or have held prominent positions in official and unofficial debates about the river and know of each other's positions and discourses.

To describe the lives of residents who work in several different Indian cities and to chart out wastewater flows within and between these cities, I engaged in a multisite ethnography (see Marcus 1986:171–73; Marcus 1995). A multisite ethnography attempts to trace the translocality that results from the fluid lines of communication between people who converse across geographic terrains and identify with each other despite their physical distance. It is an important anthropological research method for the imagined state, religious organizations transcending nation-state boundaries, the transport of ideas through cyberspace, and flows of waste. However, this study follows translocal paths that people create out of particular kinds of associations with an important river. The river is what joins one locale and one cultural group to another. Discourses about Gaṅgā are produced by those living near, worshiping, and maintaining occupational connections with the river. They are grounded in this contiguity. The discourses that I began this study with are those that are most apparent on and about the banks of the Gaṅgā.

When following an outward layering methodology and writing a multisite ethnography, the problem of boundary drawing emerged quite persistently. For instance, the informants I bring together in this book cannot be called members of an ecological population or a resource user group to conform to categories popular in ecological anthropology or the natural resource sciences (see McCay and Acheson 1987; Moran 1990; Rappaport 1990). Rappaport (1990:55) defined an ecological population as a unit of analysis that ties a population to a particular ecological base. Its members share distinctive means for maintaining a common set of material relations with other components of the ecosystem in which they together participate. The notion of resource user group, on the other hand, suggests that members of that group use a particular resource in a consumptive manner. However, the geographic and cultural fields covered by this study can conform to neither because (1) the ecological "base," if you will, is not a single entity but a series of contiguous watersheds for a transregional river, and (2) those included in this account do not engage together in identical resource use practices. In terms of geography and ecology, the river ecosystem is transregional and includes processes and systems that involve more than one watershed. Moreover, resource use practices must be considered in terms of cognitive models of ecology (what Rappaport [1984] called *cognized models* and others refer to as *ethnoecologies*), human settlement patterns, land and water uses, and forms of religious and political organization.

The population groups inhabiting my four field-cities interact with the river in a number of ways and define those ways differently. To lay out a simple difference, when performing ritual ablutions in the Gaṅgā, Hindu religious leaders, ritual specialists, and pilgrims do not think they are "using" a "natural resource." They ask the outsider to consider their engagement in other ways. For many of the Hindus I interviewed, the Gaṅgā basin is part of more than one order of symbolic and practical importance. So if Hindus in northern India argue that Gaṅgā does not occupy a "portion of the biosphere" but encompasses a cosmic order, then how should her embeddedness in ecology be treated? Or to put it in their terms, how does ecology become embedded or human materialism become contextualized in Gaṅgā's sacred order? Government officials in environmental management programs, on the other hand, encourage others to disengage from the river in the name of environmental protection. Meanwhile, industries attempt to hide their intensive extraction and utilization activities. A multisite ethnography can begin to bring these geographic, ecological, and cultural components together and facilitate the analysis of sets of data alongside one another. When layering these data sets, I am able to make assessments about how various cultural groups operate and what their forms of power are. The analysis will use these assessments to understand the tourists' complaints about pollution more evocatively.

Overview

I have built the text in such a way that the reader is taken through multiple field sites and introduced to the multiple discourses resonating in and between these sites. The book is divided into four parts so that a sense of these separate discursive layers can be communicated. But the mix of discursive traditions that each data set communicates means that these layers are not unidimensional or mutually exclusive. These discourses intersect at many angles, so a division of the book into four parts is just one way to begin looking at these angles. This organization should not be considered the only way to interrelate these layers.

Part 1 traces the contours of the geographic, cultural, and theoretical fields that I explore. In chapter 1, I position this study in the scholastic field and explain how theoretical trajectories in anthropology, sociology, religious studies, and the interdisciplinary field of political ecology provide conceptual sources for this exploration. Chapter 2 distinguishes between academic, Hindu, and official understandings of purity and pollution and then relates them to distinguishable discourses. After making these semantic distinctions, I move the focus to the unique interplay between religious, scientific, and official understandings of the river. In chapter 3, I trace out the salient components of this river ecosystem and juxtapose the scientific view of the river and its watersheds to Hindu discourses on purity and flow.

Part 2 brings together several Hindu discursive configurations that focus on the sacred river Gaṅgā. The Indian citizens I introduce use the Hindi terms for purity, impurity, cleanness, uncleanness, flow, and power to give meaning to their understanding of the confluences of wastewater with this sacred river. In chapter 4, I focus on interpretations of Hindus who reside in Banaras, a sacred city that borders the Gaṅgā. Chapter 5 explores the opposition to a canal project built by colonial officials in the early twentieth century. Religious leaders opposed colonists' attempts to divert the flow of the Gaṅgā away from the town of Haridwar and the sacred space of Har-kī-paurī ghāṭ. These parties began a debate over Gaṅgā's essence and purpose that continues today between government officials and residents.

In part 3, I explore variants of official and legal environmentalist discourses as they circulate in riparian cities and between these cities and state, national, and international capitals and centers. Beginning with official representations of the river in chapter 6, I analyze the ways in which government officials have applied the terms *defilement*, *fouling*, *contamination*, and *pollution* in water utilization schemes. I use court documents from several well water cases to show the colonial intention to define the fields of the ecclesiastic and the civil, and to guide and confuse transitions from common property and caste-based regimes of water management to private and state-directed regimes. The cases show that legal discourses, while developing against Hindu notions of caste impurity, have been crucial to the construction of a theory of natural ecology, a theory that is distinguished today from the religious approach to the river.

Chapter 7 outlines postcolonial law and litigation to explore the more contemporary sources of information for official environmental concepts. However, I explain that many of the nation's new environmental laws were not enforced until activist lawyers and citizens began submitting public interest writ petitions in the mid- to late 1980s. A new form of judicial activism emerged through the Ganga Pollution Cases. Together the cases and this form of activism make this the most significant water pollution debate in the history of Indian environmental law.

Part 4 outlines official, activists', and religious leaders' descriptions of the physicochemical and ecological dimensions of the river and wastewater flows to highlight debates over what I will call wastescapes. In chapters 8 and 9, I look at the ways treatments of these wastescapes are informed by debates over water quality and wastewater disposal and treatment technologies envisioned in India and abroad. Chapter 10 returns to a set of Hindu interpretations of waste and sacredness to begin explaining the hegemonic power of religious discourses. Profiling several Hindu religious leaders, I argue that their retreat from official notions of environmental pollution weakens the power of the Indian state and the viability of its neoliberal policies on river pollution abatement and cleanup.

The conclusion summarizes three forms of power that operate through these discourses and suggests that wastescapes themselves also constitute a formidable form of ecological power. When taken together, these layers of data widen our focus and provide insight into how people use discourse to achieve hegemonic effects, negotiate bureaucratic and legal rules, and attempt transformations in cultural and political order. In the process, these multiple layers of data shed light on the historical constitution of these salient discourses, the execution and outcomes of waste management projects, and the interplay between executive and judicial branches of the government. They also make wastescapes a visible part of the human-environment problematic.

PART I

CHAPTER 1

Contours of the Geographic, Cultural, and Theoretical Fields

THE IMPORTANCE AND CENTRALITY of water to human life cannot be overstated. There is little wonder then that rivers are a source of worry for many today. Like other valued resources, flowing rivers are now burdened by the civilizations they have watered and sustained. Human settlements across the globe use rivers as receptacles for wastewater, solid waste, and myriad contaminants, and while damming and diverting them depend upon that resource for at least half of all water consumption needs. The Gaṅgā or Ganges River in India, like the other great rivers of the world, is an important water resource and, while occupying a unique place in the history of human civilization, serves today as a waste receptacle for millions of residents of India.

Eastern and Western hemispheres have their water worshipers and romanticists. Some attribute transcendental power and others aesthetic beauty to how and where the rivers flow. These are the themes of power and wonder that underline what some consider to be a cultural paradox today. The river Gaṅgā, known to Hindus as Mother Gaṅgā, is sacred. She is a mother, goddess, purifier, and sustainer of all life. Yet to the less faithful and to followers of other religious persuasions, this important river is polluted, and in some reaches it is almost dead. What outsiders and some secular-minded Indians express is this: if Hindu citizens across India consider this river sacred and worship her as a purifier, goddess, and mother, why do they allow her to become polluted by industrial, urban, and household waste? How can firm believers in the purity of the river disregard wanton acts of pollution?

In this book, I approach this paradox as a problem for anthropological analysis. However, I do so by turning the outsider's (in this case, Westerner's) view of the problematic upside-down and exposing it for its own historical logic and contradictions. I move the inquiry beyond the questions this paradox generates to questions about the cultural construction of this paradox and the environmental concerns that inform this vision of paradox. Exploring these lines of inquiry, I try to work out the historical and discursive reasons why Indian citizens interact with the river Gaṅgā in the ways they do today.

I began this exploration by observing the human predicaments that accompany or result from the accumulation and disposal of waste, that by-product of all the consumptive practices and processes of human existence. Along the way, I was forced to think about why waste accumulation and disposal are not considered serious problems for anthropology today. Drawing on ideas from environmental activists and others I met in the field, I thought about how I could make this problematic of waste more visible to anthropologists and other social scientists. Waste includes solid material and wastewater from point (direct) and non-point (indirect) sources. In this text, I create the concept of *wastescape*, borrowing the term *scape* from the notion of landscape and annexing it to the biophysical term *waste*, to direct the reader's attention to human, industrial, agricultural, and urban wastewater flows and their impact on a sacred river and on cultural practices. The discussions that follow aim to address the following questions: What are the particular trajectories that interpretations of wastescapes take through their many practiced, political forms? What can these interpretations tell us about cultural practices that provoke and respond to ecological change? What can they tell us about ecological change?

This project bridges three foci. The first is the problem of waste disposal, an unrecognized problem in anthropology today. The second is a focus on river water resources; these are crucial resources for communities in riparian zones across the Gangetic Plain. The third is a focus on the urban setting, a topology that has received less attention in environmental anthropology than rural, natural, wilderness and global topologies have. I weave my study through an eclectic array of data sets, elicited by way of several data collection techniques, to analyze how ethnosemantic domains become embedded in interpretations and assessments of the river and inform several different discursive configurations.

As I move through this array of data sets, I ask the reader to understand the intersections of the various discourses elaborated here as more than linear movements of negotiation and contestation. The discussions that are developed tack through several configurations by way of an outward layering methodology. Beginning from a specific problem with a defined space and resident group, I trace within and outward from this space the discourses that have historically tied Indian citizens to the Gaṅgā. Reaching out through contemporary urban settings, I show that these discourses create polyphonic debates about water uses, wastewater management, and the meanings of waste and sacred power. They also shed light on the inner working of the state institutional apparatus, on relations between executive and judicial branches of government, and on the circulation of wastewater in urban, riparian zones.

The outsider's first impression is that these multiple discourses create an unwieldy picture. However, when excavating and layering pieces of them as data sets, I came across points of intersection, articulated in formal and informal meetings between government officials, scientists, lawyers, religious leaders, and citizen groups. Upon closer examination, I found that points of intersection emerged at the level of signification, as Indian citizens conversed with one another from distinctively different conceptual and symbolic starting points. I began to see that, at some times, citizens shared common terms and, at others, used these common terms to diverge along radically different discursive paths.

This book will show that studies of the ethnosemantic categories that make up forms of knowledge, traditional or otherwise, can be used in very productive ways. I use an ethnosemantic methodology to trace out how particular terms and concepts are embedded in discursive configurations over time. During the 1990s, it was fashionable to talk about discourses in the Foucauldian tradition and to characterize culture in terms of voices rather than semantic categories and cognitive structures. Foucault's notion of discourse brought attention to bear on what has been said about a subject and on the circulations of power operating through these speakable spaces. His diffuse notion of power and liberatory notion of communication were appealing to anthropologists and social scientists, especially to those who examined historical transformations in knowledge and global processes of interaction. Foucault's interest in a noneconomic analysis of power led him to the conclusion that discourses can constitute forms of power for speakers and for the topics about which people speak. Individuals and groups can use discourses to gain power for themselves or for a conceptual configuration (such as sacred purity or Mother Gaṅgā). The actual analytical understanding of power remains, however, a matter for measurement. This means that the discussion must move beyond Foucault's diffuse and imprecise caricature of power to locate the ways in which particular discourses are persuasive and operate vis-à-vis other discourses and practices.

I prefer to use Gramsci's notion of hegemony to assess the power of Hindu discourses alongside the power of official and activist discourses as they focus on the same problem: the confluences of wastewater with a sacred river.[1] In other words, I have to ask, Which assessment is more persuasive and why? How do these assessments move people to act or to legitimate their actions through this discourse? How do these assessments and discourses facilitate other forms of power through official and institutional operations?

Returning to the problem of relating semantic domains to discourses, I take direction from earlier work on cognitive structures and semantic categories.[2] Although critics argue that these older approaches have reified thought and language to a realm autonomous from practice, the material conditions of production, and ecological processes (see Burling 1964; Harris 1968; Vayda and Walters 1999), I will argue that these accusations are not entirely fair. Many scholars have related cognized models, ethnoecologies, or traditional ecological knowledge to other levels of cultural practice and to ecological systems, processes, and events.[3] Like Rappaport, many have tried not to treat cognized models as stand-alone constructs (see Maffi 2001). Wolf reminds us that Rappaport (1984:238–39) defined a cognized model as the model of the environment conceived by the people who act in it and who interpret the fundamental relationships of what they consider to be the environment. Rappaport related cognized models to operational models—those models created by the anthropologist through observation and measurement of empirical entities, events, and material relationships—but did not consider the two isomorphic in any necessary way (Wolf 1999:19). Cognized models help the analyst understand the motivations for human behavior and resource use, while operational models relate behaviors and uses to ecological cycles and processes objectified by the analyst.

Critics have also argued that models of native ethnoecology or traditional ecological knowledge have separated culture from the decision-making process and privileged culture over ecological events, processes and systems. Ellen (1982:209–26) has argued that studies of native classifications of soils, topography, and plant and animal life have tended to project static models without reference to place or social context. My general interest here is to show how ethnosemantic categories become central to discursive configurations that form multilayered paths of communication between different cultural groups. Taking this focus on discourse, I hope to get at the intersections of varied discursive units while also tracing the paths to their respective focus centers. These crossroads are the points of articulation between competing, opposed, or divided groups, while the centers of each discourse tend to include resistances against other discourses.

The methodology of ethnosemantics does present a problem, however, and this lies in its underlying notion of sharedness. Studies of native classifications have tended to invoke assumptions of sharedness at the level of linguistic terms and their meanings, to promote the view that those who use the same lexemes share the same culture.[4] There are two problems with this kind of boundary drawing. First, an analysis of semantic domains or ecological knowledges may miss the contested way in which domains are entitled through social relationships. Semantic domains shift as speakers debate with one another and share experiences that impact their daily lives. I use the methodologies developed by ethnoscientists to explore multiple interpretations of the Gaṅgā and wastewater by pointing attention to the complexity of the linguistic and conceptual realms.[5] Then I relate the linguistic and conceptual realms to human intentions, cultural practices, institutional rules and outcomes, and wastescapes through an outward layering methodology.

Extending Foucault's notion of governmentality to the sphere of state resource management practices, Brosius (1999b), Gupta (1998), and Luke (1999) have traced out regimes of resource or environmental managerialism at local, national, and global levels. I will show through this ethnographic excursion that environmental governmentality (what Luke calls *environmentality*) must be broken into its historical components and related to other dominant discourses over time. The governmentality I outline here results from a series of explicit and implicit bureaucratic and legal rules that have formed layers of discourse and practice. Government officials manipulate bureaucratic and legal rules to generate personal gains. So governmentality in this case is not about generating a viable hegemony, a persuasive code that the public subscribes to; rather it is about using discursive positions to master layers of implicit and explicit rules and negotiate a way through them.

Looking at other studies of citizen-state interactions in South Asia, I take my lead from Brass (1997), who has provided an instructive analysis of the discourse on riots in northern India. He introduces the concept of intersecting discourses to convey fields of voices that do not have stable connections to particular interests, frameworks of knowledge, or cultural practices. His model introduces a level of shiftiness to the whole enterprise of signification and communication and to the ways signification and communication are put into domination and exercises in truth making. I will refer to his ideas throughout the book.

Returning to the problem of Hindu conceptual domains, I want to orient the reader to the conceptual complexity of the Hindi terms for purity, impurity, cleanness, and uncleanness while not positing them as unchanging or undisputed frames. When excavating these data sets, I have found that familiar terms invoke different conceptual configurations. An analysis must trace how speakers use key lexemes as shared points of departure for divergent discourses.

Let me lay out a preliminary example. When Hindus and scientists use the word *pollution* or the Hindi transliteration *praduṣan*, they are generally invoking as well as reinventing different conceptual constructions. These conceptual constructions are also, in some cases, different from the environmentalist's and the anthropologist's notions of pollution and impurity. Residents of the field sites I cover in this text revealed this conceptual pluralism when using the term *pollution* to talk about waste and the people who manage it. Transliterating the English term *pollution* in Hindi as *praduṣan*, residents of Daśāśvamedha associated a person rather than an ecological process with the condition of pollution. They called officials in sewage management offices *praduṣanwālās* to make reference to those who operate with, do, or are pollution. In the process, they defined the very nature of what is meant by the word *pollution.* Defining a person by the focus of his or her occupation, they used personhood and moral demeanor to comment on what I saw as an environmental problem.

This labeling follows the procedure of naming individuals by and as the objects of their work. There are *pānwālās* ("those who sell pān"; literally "one of pān"), *rikśewālās* ("those who pedal rickshaws" or "those of rickshaws"), and so forth.[6] By metonymic association, Hindi speakers in these cities accused humans rather than the "natural" world or the ecosystem for the problems they addressed, and by extension thought of material pollution in direct relationship to the people who created it. They did not distance or essentialize pollution as an ecological state. When personalized, however, this title brought on a status different from that ascribed to low-caste workers who deal in scavenging, recycling, and the cleaning of streets and sewers. The latter are scapegoats in a different way (see Korom 1998). Initial efforts to understand the various meanings of *pollution* made this a problem of deciphering layers of polyvalence.

The problem of polyvalence is central to the process of articulation between discourses. The anthropological notion of polyvalence or multivocality, developed in the work of Turner (1975), Fernandez (1974, 1977, 1998), Ohnuki-Tierney (1990a, 1990b), and others, follows from Geertz's (1973) notion that symbols are public vehicles for conceptions. Leach (1990:229) has noted that a polyvalent symbol provides a tangible relic for the construction of myth and ritual. Sahlins (1976) has argued that multivocal or polyvalent symbols draw attention to *a dynamic center* of signification enterprises. To use Herzfeld's (1993:26) phrase, polyvalent symbols are semantic vessels filled with a variety of ideological messages. When symbols are put to the multiple readings of a diverse audience, to readings that differ by culture and by interpretive license and creativity, and when they retain a dominant position, an effective polyvalency is achieved (Ohnuki-Tierney 1990a:17). As Handelman and Shamgar-Handelman (1990:217) have put it, few symbols speak for themselves without immanent instrumental referents or functions. These referents and functions are powerful through their polyvalency. These polyvalent intersections unite audiences and take them metaphorically to uncharted landscapes.[7]

Since this discussion turns attention to discourses that, in Foucault's words, partially create the topics about which speakers speak, the challenge is to retain the understanding of ethnosemantic categories and polyvalency alongside considerations of discursive configuration. By considering linguistic terms as public, polyvalent symbols, I will show how the concepts signified by common sound images give landmark definition to several distinctive popular discourses. The common terms *purity*, *impurity*, *cleanness*, *uncleanness*, and *pollution* become points of discussion for various cultural groups. They are polyvalent intersections that bring to light the multilayered nature of communication.

Waste Politics and Social Science

Moving along with my outward layering methology, I found that a parallel process of theoretical integration was required. How should these instances of discursive centering, that peel away into separate cultural spheres, be related to institutional practices and policies and to official or industrial achievements or failures? I looked to social science models for guidance in this integration but found none that communicated a comparable problematic. It is unfortunate that anthropologists have not taken up in a rigorous manner the study of the politics of waste at local, national, or global levels. Perhaps it is the culturally neutral, unexotic appearance of waste flows that keeps them at bay. Or have they been hidden from cultural critique because of their grotesque forms and their smelly, dangerous, and downright degenerative qualities?

Though not viable as models, a few works stand out as notable exceptions to this deficiency. Several anthropologists, taking their lead from Douglas (1985, 1992) and Douglas and Wildavsky (1982), have discussed waste and pollution in terms of the cultural construction of risk.[8] In her study of community responses to the toxic contamination of groundwater, Fitchen (1987, 1989) outlined how New York residents used the cultural construct of home to shape interpretations of the risks of pollution. Kottak and his colleagues (Costa et al. 1994; Kottak and Costa 1993) undertook on a multisite study of risk perception in Brazil, starting with the risks of nuclear energy and expanding the focus to risks associated with deforestation, industrial pollution, and pollution from mining and public sanitation. But, these studies aside, attention given to waste politics has been rather meager.[9] This is all very unfortunate given the crucial roles that waste production and management play in cultural life and local and global ecosystems today.

Critiques of development, capitalism, and transnational commodity trade have also ignored flows of solid waste and wastewater and have given only limited attention to transfers in waste treatment technology. Even scholars concerned with governmentality in the field of the environment have not considered waste disposal and treatment (see Brosius 1999a, 1999b; Escobar 1999; Gupta 1998; Luke 1995). Analysts of the city, a space of increasing importance for global capitalism, have also underreported in the areas of waste management and water and sewerage services (see Low 1996).

Outside anthropology, the climate is different. Geographers, sociologists, journalists, and public policy makers have been more concerned with waste politics. Using actor-oriented, decision-making approaches, many have focused on the interpretations of risk made by waste industry personnel, government officials, and residents of communities sited for waste facilities.[10] Others have looked at the ways in which gender roles and images influence grassroots resistance to waste management practices (Bru-Bistuer 1996; Miller et al. 1996). Beder's work on the sewage system of Sydney is the best example in the field of science and technology studies. Beder critically explores the historical and contemporary politics of sewage management in Sydney, in terms of changes in technologies over time and their uses by key institutional players.[11]

Taking a macroapproach to the study of organizations and movements, environmental sociologists have used two models to analyze organizational structures in what they call the environmental movement. Some have described how collective action emerges from collective perceptions of identified problems.[12] Theories generated by this work suggest that social movements develop as increasing numbers of people become frustrated with a specified set of conditions, recognize a common discontent, and agree on a plan to change these conditions. This has led to the assumption that social movements have a natural history or life cycle that culminates at the stage of institutionalization (Bakuniak and Nowak 1987; Blumer 1939; Smelser 1963; Turner and Killian 1972). Then, after institutionalization occurs, the original enthusiasm and idealism withers away and bureaucratic decision making takes over. Borrelli (1987), Commoner (1987), and Sale (1986) have used this model to outline the evolution of the environmental movement in the United States.[13]

The alternative "resource-mobilization" model focuses on the procedures organizations follow to promote their objectives (McCarthy and Zald 1973, 1977; Tilly 1978; and Turner 1981). The resource-mobilization approach identifies the main sources of support for group action and characterizes the relationship between a social movement and its social context in symbiotic rather than adversarial terms. In environmental sociology, the life cycle and resource mobilization theories have provided frameworks for investigating the formation and development of environmental movements.[14]

Culture and Ecology

While contributions from environmental sociology are worth noting, it is my opinion that models in the multidisciplinary field of political ecology provide the strongest starting point for building an approach to the politics of waste disposal and treatment in India. Studies in political ecology draw attention to the political and economic appropriations of natural resources and to the effect these appropriations have on ecological processes.[15] Recent studies in feminist political ecology have investigated the intersections between gender and science, property and resource rights, and political activities and environmental concerns (Rocheleau et al. 1996).

The problem is that political ecology is a field of some controversy. Part of the controversy is tied to old dualisms that have haunted social science for some time, dualisms that hinge upon a fundamental epistemological discrepancy over what can be known through scientific inquiry. Simply put, these dualisms pit quantitative analyses against qualitative or interpretive analyses, scientific approaches against humanistic ones, evolutionary views against historical frameworks, and biophysical processes against cultural structures and individual practices (Crumley 1998:xi). Even if these dualisms do not totally frame theoretical debates, they nag at the consciousness, forcing scholars to find their positions between or beyond them and threatening to reduce very complicated subjects of inquiry to sets of opposing variables.

This tension is evident in debates between cultural ecologists and political ecologists, and more recently between political ecologists and poststructuralist political ecologists (see Escobar 1999; Soule and Lease 1995). In earlier debates, cultural ecologists argued that political ecologists were neglecting to analyze both sides of the human-ecology dialectic, and political ecologists were leveling the same charge at cultural ecologists (see Escobar 1996, 1999; Kottak 1999; Peet and Watts 1996; Rappaport 1990; Vayda and Walters 1999). While cultural ecologists were considering cultural practices within the rubric of an ecological order, political ecologists were subordinating ecology or nature to resource competition and to class relations in a political-economic structure. The former group has looked at small-scale societies, while the latter has situated simple and complex societies within a global world system of development and underdevelopment. Generally cultural ecologists have exhibited a deeper understanding of the natural sciences, while political ecologists tend to draw from a range of debates in Marxism, development studies, political science, and economics.

More recent debates between political ecologists and those adding a poststructuralist or constructivist orientation have exhibited similar tensions. When examining social movements, development activities, and environmentalism, poststructuralists have included a serious focus on problems of language and discourse (see Brosius 1999a, 1999b; Escobar 1996; Peet and Watts 1996). This direction combines an appreciation for meaning, language, and discourse with a focus on power, production, natural resource use, and ecological change, often in a global context. When poststructuralists posed the constructed and discursive nature of *all* cultural and ecological paradigms, to desettle the assumption that truth is lodged in a natural order, political ecologists rebuked this constructivism, arguing that ecosystems can be and should be described with a degree of scientific certainty and objectivity. Opposing notions of context and causality—humanity in ecology or ecology in humanity—continue to underlie these arguments.

Amid this, there are scholars and scholarly working groups who are attempting to transcend this impasse, by synthesizing models or levels as Rappaport had done much earlier (see Wolf 1999). For example, historical ecologists aim to use all viable units of analysis at their disposal to approach a problem: science and humanism, evolutionary and historical frames, and quantitative and qualitative methodologies (see Balee 1998a). Others sympathetic to the deconstructionist approach look for ways to use the concepts of interactivity and positionality in a "constrained program of deconstruction" (Hayles 1995; see also Alcoff 1994).

I use the outward layering methodology to bridge considerations of discourses together with considerations of cultural institutions and practices and flows of wastewater. Studying the ways these data sets articulate cannot be guided by a dualistic debate about whether it is culture or ecology that plays the dominant role. As I see it, both culture and ecology are so deeply intertwined in and between layers that it is almost pointless to separate them into opposing forces.

Moving away from a focus on wholeness and system stability, I aim to show how a number of cultural groups interpret the transformations produced by something more marginal to culture and ecology: the flows of wastewater that fill the interstices of everyday life. As marginal substances, wastewater flows cause transformations in ecological and cultural systems and processes as they progress along diffuse paths. In serious cases, wastewater flows can outpace the regenerative power of ecological systems and stretch the resiliency of the human immune system. A focus on wastewater flows from this perspective must include analyses of the production processes (household, agricultural, and industrial) that produce wastewater, the infrastructural processes that guide its flow outward and across space, the urban or built environments through which wastewater flows, and the ecological processes that produce its basic ingredients and that come about as wastewater diffuses through a river basin. To understand the relations between these various layers, however, we must turn to the ethnographic discussion.

I do not begin with the assumption that an environmental movement against river pollution has emerged in India. Instead, I describe a field of intersecting discourses that, in their multiplicity, undermine the possibility of a coherent and unified movement.[16] In discussions about the river Gaṅgā, there is marked disagreement about what purity, impurity, cleanness, uncleanness, and pollution mean. Among those who speak about these meanings and essences, it is impossible to find a singular collective consciousness or a unidirectional form of action emerging from that consciousness.

I also use this study of multiple discourses to look at the inner functioning of executive and legal institutions and to assess the relative power of various cultural groups to manage or control this crucial water resource. The outward layering methodology brought me into the wider discourses that stretch across government agencies and industrial plants and that circulate common ideas about the scientific assessment and secular administration of resources. In India, urban pollution prevention is not a modern phenomenon. But current schemes differ in scope and quality from colonial and precolonial ones. The latest wave of official attention has developed alongside the liberalization of the Indian economy. Drawing for legitimacy from the substantial amendments made to the Indian Constitution in the 1970s, Rajiv Gandhi defined environmental pollution as a national problem early in his tenure as prime minister. In 1986, he created the Ganga Action Plan and led official bureaucracy into a new phase of administration.

The outward layering methodology also took me into the colonial past, where I layered these colonial data sets alongside data from the postcolonial period. Official discourses today continue a debate over the essence of the Gaṅgā that started with legal arguments over well water and with the assessment of defilement and caste impurity. They also respond to more contemporary international debates and policies. My study, like many others focusing on contemporary India, grows out of a concern for the postcolonial condition (see Apffel-Marglin 1992, 1994; Apffel-Marglin and Marglin 1990; Breckenridge and van der Veer 1993; Gupta 1998). It is therefore attentive to the ways in which contemporary local-national-global articulations are related to colonial policies, laws, and bureaucratic structures and to a postcolonial remaking of them.[17] At the same time, my critique abandons the top-down concept of power engendered by development paradigms to show that power may occur at many discursive and institutional levels today and that players may shift voices as they move through different discourses. The local is not merely subordinated to the national or global, nor, as this text reveals, are religious discourses subordinated to official or activist configurations.

Entitlements

Escobar (1996:326) has argued that transnational exchanges follow a new process of capitalization, a process that M. O'Connor (1993) dubbed the *ecological phase.* In this phase, nature is not defined and treated as an external, exploitable domain (as it was in the phase of pioneer capitalism); rather it becomes a commodity, managed and conserved for greater consumption. This phase coincides with what J. O'Connor (1988, 1989, 1992) has called the second contradiction of capitalism. The second contradiction occurs when stakeholders encroach upon nature so deeply that they impair conditions of production and resource availability.

Taking Escobar's and M. O'Connor's definition, one might find the ecological phase resonating in the voices of U.S. administration agents as they debate global warming and climate change.[18] In conferences focusing on this global problem, participants from the United States have sought to maximize their scope for industrial and technological advancement in a world in which heavy legitimations for these pursuits are required. Attempts are made by players from advanced industrial countries to use environmental conservation and protection rhetoric as a way to mediate their compliance to international agreements and treaties. Paradoxically, national leaders use legalistic, bureaucratic language and rhetoric to justify their claims to manage and "conserve" these resources, for ever-greater levels of consumption (see Brosius 1999b).

National debates on pollution prevention in India show signs of a push toward what Escobar and M. O'Connor have called the ecological phase, but the movement is dominated less by technological and official interests and more by the privatized agendas of NGOs and environmental activists. On the other hand, the broader public is engaged in debates of a very varied sort. In resource competition, we find a dynamic tug-of-war between many groups—property owners, commons users, regulators, conservationists, religious leaders, devotees, and the like. These tugs-of-war occur at the crossroads of religious faiths and practices, nationalist movements, judicial activism, and popular culture and suggest that a crucial phase in the symbolic contestation of rivers is under way in India. It is a discursive phase in which industrialists are less visible and less skilled in ways of using conservationist language to legitimate their interests. Meanwhile, some bureaucrats, NGOs, and activists employ conservation rhetoric quite fluently to advance their particular agendas. But discourses are not just framed by conservationist or protectionist ideas. They are also tied up with notions of the sacred and the transcendent and with assessments of the micro- and macropolitical economies of waste disposal and treatment.

Instead, debates about wastewater and the river Gaṅgā in India appear to elude Escobar's ecological phase as they involve disputes over who is allowed or entitled to symbolically define the river. These entitlement claims are played out in the disciplinary and cultural fields of politics, religion, science, and law. These claims are made in debates that have no single center. Although some arguments appear close on occasion—when an activist, a lawyer, and a pilgrim priest defend a devout Hindu's right to bathe in the Gaṅgā and undertake purification rituals with her water—they are in fact only intersecting before moving out along separate semantic paths.[19]

By examining debates about the character and official and unofficial management of river water resources, I repeat that my aim is not to deflect attention away from the activities of industries and other official players who dump effluent in this waterway. In fact, by tracing debates as they are shaped through everyday practices, religious rituals, and the activities of research, policy-making, and litigation, I argue that some Indian citizens are not involved in debating how the Gaṅgā *should be* used or worshiped. Generally, industrialists are absent from these debates because they want to take a more invisible position and hide or mystify their own polluting behavior. They do not incorporate conservationist or protectionist notions into their own policies. However, pollution problems do not stem from industrial uses alone; they are created and facilitated by official and public conceptions, discourses, and practices.

The polyphonic nature of debates about the Gaṅgā is also a sociological result of river ecology. Unlike the contested frontiers of Amazonia (see Schmink and Wood 1992) or the submergence zone of the Sardar Sarovar dam (see Baviskar 1995), the river is not disputed in terms of ownership, except in terms of ownership of flow quantities. When ownership of flow quantities is debated, however, it is only at the highest bureaucratic levels, within the Central Water Commission and between the Indian and Bangladeshi governments in treaties over water use rights. These debates do not involve private citizens. Instead of focusing on water rights and flow rights, residents of settlements in the river basin tend to focus on the manner in which the Gaṅgā is defined in public policies and religious traditions and on the ways in which local forms of consumption and worship should be understood and regulated. These contests presuppose the fact that Gaṅgā's waters are not and can not be controlled or bought by a single cultural group or network. This sharing feeds disputes about proper ways citizens should behave as guardians, worshipers, appropriators, and exploiters.[20] Therefore, a focus on this field should include the agency of industrial players and the power of production in the wider, more layered fields that many members of society participate in, to reach into the decisions that shape specific human connections to this river.[21]

Rivers are generally left out of studies of common property resource management, which focus instead on irrigation systems, smaller water bodies, and forest tracts as targets of resource regimes in various stages of articulation with other regimes (Berkes 1985; Blair 1996; Bromley 1992; Feeny et al. 1992; Lansing 1991; McCay and Acheson 1987; McGrath et al. 1993).[22] When attention has been paid to rivers, the focus has centered on water rights (Derman 1998; Greaves 1998) and contests over dams and other forms of flow management (Baviskar 1995; Ettenger 1998; Loker 1998; Rogers 1998; Sheridan 1998). Wastewater effects on rivers have received little attention in this emerging interest in river water resources.[23]

The data show that the definition of open-access resource is a problematic one, a point that many have underscored by neglecting to use it, because river resources are both open and regulated and used by cultural groups of various sorts. The Gaṅgā is open to all (in certain locations and at certain times), yet managed (not owned), in both physical and rhetorical senses, by those vested with legal right and responsibility and by those endowed with sacred authority. These players use conflicting codes to carve out legitimate places for themselves in public discourses. This means that "management" is part of a competitive realm, and only partially coopted by a state vested with legal authority. Beyond official management, codes in religious texts and ritual discourses and practices act as the more persuasive blueprint for public use of the resource. Hindu religious leaders and ritual specialists insist that Gaṅgā's purificatory power and motherliness cannot be reduced to the materialist logic of scientific pollution. Their aim is to teach pilgrims how to worship Gaṅgā as a divine power. They argue that official policy cannot define the meanings of practices of immersion (what they call *visārjan*). In the Hindu code of worship, statues *(mūrti)* of gods and goddesses, religious books, and saris are offered to Gaṅgā at certain moments of cosmic juncture and fall outside the category of waste defined by law and official policy.

Hindu discourses and rituals of purification have guided cultural approaches to the river for centuries. Before colonial schemes took a more instrumental view of the river, those who worshiped Gaṅgā and practiced rituals of ablution commanded considerable respect among the population and respect among non-Hindu rulers as well. These ritual specialists and practitioners continue to hold the power to contest, at a grassroots level, ideas about river pollution and natural ecology. At times, religious leaders and practitioners clearly distinguish their ideas from ideas about science, ecological systems, and a universe without divine power. At other times, they consciously embed scientific, ecological, and mundane power in sacred power and in religious narratives and practices.

Academic scholarship has distinguished the Indian nation-state as a democracy bound up with religious concerns and practices (see, for example, Madan 1987, 1993; van der Veer 1987, 1994, 1996). Hindu nationalist movements and the activities of other secular political parties have clearly demonstrated how religious ritual practices become instruments for political mobilization and interest-based politics (see Hansen 1999). This means that, as many South Asianists would emphasize, religion is a hegemonic force, casting a shadow over attempts by state officials to create a secular understanding of modernity in current nationalist discourse.[24] It follows that religious leaders, ritual specialists, and the organizations *(saṃpradāy)* or sects they represent cannot be subordinated to the power of the executive, legislative, or judicial institutions of the nation-state, as if the political jurisdiction ruled over the religious. Religious traditions and institutions claim to be forces in their own right; they guide and oversee the spiritual convictions and occupational positions of large sectors of the Indian population.

The Break between Discourse and Practice

Looking back then to my general focus on discourse, I would like to emphasize the point that I chose this focus not because of my interest in poststructuralist theory but because of the urgency apparent in the remarks of my informants. Informants often made it clear to me that environmental programs to save, protect, or clean Gaṅgā, whether at the official or grassroots level, have produced little more than talk or speeches (*bātcīt* or *bāṣan*). No one has suggested that a true movement to stop pollution has emerged from the words on people's lips. But the emphasis my informants have given to talk and speeches does not preclude attention to behavior, politics, natural resource use, waste management schemes, or ecological systems, processes, or events. Rather, their insights point to the ways various schemes to save, protect, or clean the river demonstrate other concerns and promote other practices.

Their focus on discourse is also linked to the argument that because words are just as elusive as behaviors are, it may not be necessary to connect a specific discourse (or the allegations within it) to a set of practices that appear to correspond to it (see also Brass 1997). This reflects their understanding of the break between discourse and practice, an understanding I needed to take as a precondition for investigating this cultural complex, not as a theoretical endpoint. Elusive words and behaviors are also evident in the discourses and practices of industrialists and government officials. So the waste disposal practices of industry and the waste treatment practices and policies of government agencies had to be interrogated in other ways, first through discourse and then through other links implied by the discursive positions of key players. So instead of breaking into industrial or wastewater treatment plants in a militant fashion, I found it more productive to collect stories I heard about bribes taken for positive inspection reports, about death threats made to inquiring journalists and academics, and about the physical harassment of environmental lawyers and activists. Reading them as ways citizens voiced their resistance to a development tyranny that entices and then debilitates them, I got closer to the elusive depths of official and industrial practices. Few informants knew how to, as they put it, "bell the cat," but many knew what that cat was about.[25]

Without breaking in, I was also able to photograph and videotape emissions from the periphery of industrial and official plants and pretend, in a sense, to know what went on inside them. Doing a virtual documentation of development at the margins, I recorded emissions from corporeal edifices and charted household, urban, and industrial wastewater flows and their intersections with cultural landscapes and practices. This is a problematic that signals us to focus on the peripheries of capitalism, development, and transnational flows. It has led me to new fields of cultural and ecological data, fields I could access only after accepting a break in the correspondences between discourses and practices and enduring a persistent rudeness to the senses.

CHAPTER 2

The Polyvalency of Purity and Pollution

IN 1986, PRIME MINISTER RAJIV GANDHI created the Ganga Action Plan, the government's first environmental program. The plan aimed to prevent pollution in the river Gaṅgā and in urban centers along the riverbank. Speaking in the sacred city of Banaras, the prime minister inaugurated the program with the following words:

> The purity *(pavitratā)* of the Gaṅgā has never been in doubt. Yet we have allowed the river to become polluted *(gandā hone de rahī hai)*, a river that is the symbol of our spirituality. The felling of trees has caused severe floods and silt and mud now flow into the Gaṅgā making the river shallow so that boats cannot ply in it as they did before. Along with this we are seeing that the different kinds of pollution *(polyshan)*—the dirt *(gand)* of the city, of industry, of factories and of dead animals, all kinds of dirtiness *(gandagī)*—we are throwing in the Gaṅgā. From now on, we shall put a stop to all this. We shall see that the waters of the Gaṅgā become completely clean *(bilkul sāf)* once again.[1]

In this speech, Rajiv Gandhi made a significant statement about the nature and quality of the river Gaṅgā, cutting through several cultural logics and tying common terms together to create a curious bricolage. He was calling for a new conceptual space by setting up metaphoric linkages between categories considered distinct in popular discourse, especially in the popular discourses of scientists and Hindus. By talking about Gaṅgā's sacred purity (what Hindus refer to as *pavitratā*) and human-generated pollution (what Hindus call *gandagī* and he called the dirt [*gand*] of the city), he modified and gave new meaning to the terms for purity *(pavitratā)*, dirt *(gandagī)*, and cleanness *(sāf honā)* by juxtaposing them in the same discussion. In other words, he created another set of meanings for familiar terms.

Rajiv Gandhi set out a way to tackle an enormous semantic problem but in the process created the conditions by which government officials could talk past Hindu residents in discussions about this sacred river. The problem of semantic polyvalency is one that many government officials have ignored since Gandhi's inaugural address. Even those officials who privately practice Hindu rituals of purification with faith and devotion do not address these meanings in policy debates.

Ganga Action Plan policies often use the English term *pollution* to define a series of problems that involve waste and its impact on the river. Since the term *pollution* can gloss over a distinction between the toxic waste of industry and the organic waste of households and human interactions, it has the potential to obfuscate the understanding of ecological processes that may be harmful to residents. Such obfuscation is evident in government documents on water quality that underreport levels of toxic waste in the river by focusing on the parameters of dissolved oxygen and biological oxygen demand (BOD)—parameters that say little about specific heavy metals and toxins in the river.[2] This is problematic for the development of the public understanding of industrial waste and wastewater in general.

The English term *pollution* also confuses or disregards the distinction that many Hindus make between impurity *(apavitratā/asuddhatā)* and material uncleanness and dirtiness *(gandagī/asvacchtā)* (see chap. 4). Rajiv Gandhi did not ignore this distinction but used it to argue for an ecological understanding of the river. He used terms that were most familiar to his audience when he opened with the argument that Gaṅgā's purity *(pavitratā)* was not in question. Then he roped in the audience for a more difficult stretch by suggesting that Gaṅgā could be adversely impacted by dirt and waste *(gandagī)* and rendered unclean *(gandā)*. Arguing that Gaṅgā "needed to be clean again" *(dobārā bilkul sāf hone lag jae)*, he opened up a discursive space for considering sacred purity and impurity vis-à-vis uncleanness and ecological pollution.

Gandhi argued that Gaṅgā's purity was not in doubt. But he asked people to recognize that human activities were causing harmful pollution, transliterating the term from English to Hindi as *polyshan.* He promoted the idea that humans had the power to stop waste disposal practices that might be harmful to the river. With the power of human agency, Rajiv Gandhi explained, the Gaṅgā could "become clean once again."

The primacy he gave to human agency over ecological, natural, or divine power differs in significant ways from assumptions about Gaṅgā's power that members of his audience were likely making at the time. To deliver his address, Rajiv Gandhi lectured from a platform on Rājendra Prāsad ghāṭ, at a spot located in the center of the sacred city of Banaras and directly on the bank of the river. This flight of steps gives the public access to the river and is adjacent to the ghāṭ on which I have conducted my fieldwork. He spoke to a modest crowd of residents who had gathered on the riverbank or were seated in decorated boats. These were residents with very different views about the power of the river.

The contemporary discussions described in chapter 4 make several things clear. First, residents of this neighborhood of Banaras generally do not connect river cleanup work with modern civic responsibility. Second, they do not subordinate divine power to human material pursuits. These are two features that underlie their view of waste and its relationship with the Gaṅgā. Rajiv Gandhi's sense of human agency reflected the motives for his work and for official-bureaucratic programs on environmental management in general. Discourse supporting official water supply and sewage disposal programs projects this image of human agency and develops methods for measuring ecological processes.

The sources of this polyvalency do not end here. Notions of purity and impurity are also articulated in the academic discourses of anthropology, sociology, and religious studies. These notions have long and controversial histories of their own. In *Homo Hierarchicus* (1980), Dumont described Hindu society as a hierarchy dominated and organized by the cultural principles of purity and pollution. Purity, as Dumont put it, is a force that encompasses social life as "the mantle of Our Lady of Mercy," and it determines in very rigid ways the nature of social interactions and interactions between people and natural resources. After the publication of this text, many scholars lined up against Dumont's view and charged that the purity-pollution principle shaded out other more important linguistic and conceptual concerns in Hindu life (see Das 1982; Feldhaus 1995; Marglin 1977). Also at this time but outside the context of South Asian studies, anthropologists were thinking about the concepts of purity and pollution in terms of cognitive and symbolic forms of classification and positing that they had an important role to play in giving "unity" to rather "untidy" everyday experiences. In these accounts, the term *pollution* was used to describe the cultural conditions, processes, and interactions that were relegated to the marginal, anomalous, and unclassifiable. Some of these polluting substances and conditions posed threats to given cultural orders. In their scholastic incarnations, the terms *impurity* and *pollution* have carried a heavy load culturally and conceptually; in religious, structural, or symbolic ways they have helped scholars cast an organizing net over society and culture.

The problem here is that anthropological constructions of purity and pollution have tended to recast emic constructs in sociological frames and reduce definitions of the transcendent to an anthropocentric base. To address this problem, anthropological notions of ritual purity and pollution must be separated from the emic ones I propose here, then understood in relation to each other and in terms of the intersections among distinct and sometimes competing discourses.

To do this, the discussion must begin with a review of the theoretical arguments in anthropology and religious studies that have developed anthropocentric meanings for the terms *ritual purity*, *impurity*, and *pollution*. This will flesh out anthropological and social scientific discourses. Since the anthropological frame can never be commensurate with the native frame it examines, identifying the conceptual baggage that accompanies central scientific terms can help to clear the interpretive field of weighty preconditions. Even so, while we aim for a comfortable fit between our own interpretations and those of the people we represent, Clifford (1986) reminds us that all we can hope for are partial truths. This partiality of truth underlines the fact that our elementary standards of measurement are always and inevitably skewed. In the measuring process, the interpretive field is rarely leveled. Given these predicaments, it is helpful to consider that as researchers we are engaging in a conversation with the frames with which we seek commensurability. The translation from anthropology to other discourses and back again is part of a larger dialogue that must be identified and situated (see Marcus and Fischer 1986:30–31).

In anthropology, sociology, and religious studies, most accounts of South Asian Hinduism have considered ritual purity and impurity as the two ends of an organizing axis of caste hierarchy. Scholars in these disciplines have argued that the principle of purity and pollution encompasses mundane matters of hygiene and physical cleanness and, by doing so, collapses them into the cultural category of ritual purity. I will demonstrate how this process has come about. A second problem is that the term *pollution* has been used interchangeably with the term *impurity*. This scholarly practice has led to a blurring of the boundaries of a distinction that Hindus make between ritual impurity and material uncleanness. To demonstrate this, I review what scholars have said about the Hindu concepts of purity, impurity, and pollution. Then I analyze layers of data to discuss the distinctions that informants and historical figures have made between spiritual purity/impurity and physical cleanness/uncleanness (see chaps. 4, 5, and 10). I relate conceptual distinctions to discourses and practices in several different fields of knowledge and power—to religion, politics, science, and law.

Purity and Impurity in *Homo Hierarchicus*

In South Asian anthropology, ritual purity and impurity are related to caste structure *(jajmānī)* relations. In these approaches, purity and impurity flow through bodily substance and become identity markers in ritual exchanges. Most anthropologists and sociologists cite Dumont's *Homo Hierarchicus* (1980) as the authoritative account of the sociological nature of this principle of purity and impurity in Hindu caste *(jāti)* relationships. But they all tend to move beyond Dumont's work by exploring other linguistic and cultural axes that underlie Hindu ritual and cultural life.[3] Several critiques of Dumont's work have argued for a more fluid and substantive understanding of purity and impurity, to move beyond the limits of a structural approach.[4] Yet, ironically, many of these critiques have reproduced the categories of purity and impurity Dumont outlined in his seminal text. While Dumont's principle of hierarchy was dismantled, his definitions of purity and impurity remained unchallenged. Therefore, it would be important to return to Dumont's summaries of purity and impurity for a moment and to the ways in which anthropologists and sociologists have reproduced these categories in their work. This will allow us to distinguish academic notions of purity and impurity from those articulated in Hindu, scientific, and official discourses.

In *Homo Hierarchicus*, Dumont argued that the principle of purity and impurity organizes hierarchy in Hindu society. For him, purity and impurity were religious notions that dealt with status rather than political power. He explained that the opposition operates through categories of people and through the regulation of interactions between people and categories. The opposition also underlies rules for defining food and its consumption and for regulating rites of passage such as birth, marriage, and death.

Dumont (1980:47) wrote that the principle of impurity, though invoking the notion of hygiene in the Western mind, should not be confused as a secular or atheistic prescription. It should not be reduced to a classification concerned with ordering mundane reality:

> We have endeavoured to reconstruct an idea which is fundamental and hence extremely all-embracing or encompassing for the Hindu. To do this, we have refused to confuse it with our habitual ideas, for example, to trace purity back to hygiene. We have isolated a predominant idea which is absolutely different from our own. This was indispensable for the understanding of the social system, but it was only the first step in the comparison of Western and Hindu ideologies, and a brief indication can be given of how this approach could be developed. A situation similar to that encountered in passing from one language to another is involved, when it is recognized that a given semantic field is divided differently. Edible plants are apt to be classified in English into fruits and vegetables, whilst Tamil opposes *kay* (green fruit which has to be cooked or prepared for eating) and *paLam* (ripe fruit). But in our case there is a hierarchical relation between different levels. The notion of purity is rather like an immense umbrella, or as we shall say the mantle of Our Lady of Mercy, sheltering all sorts of things which we distinguish and which the Hindu himself does not confuse in all situations. It is as if different configurations of notions cover the same sector of the semantic universe. If need be one could speak of function in order to

> pass from one case to the other, say for example, that the idea of purity has hygienic functions, but clearly this would be to fall back into sociocentricity. Let us confine ourselves to noting certain overlaps. (1980:60)

Dumont warned against the tendency to reduce Hindu notions of purity and impurity to Western notions of hygiene, and raised the problem of commensurability when asking scholars to avoid habitual associations. By arguing for a translation of Hindu difference that is commensurate with the native view, he orientalized the notions of purity and impurity, giving them an "encompassing" nature and distancing them from the manipulations of political power. This kind of orientalist imaging has been discussed elsewhere in terms of colonial and postcolonial practices of domination (Breckenridge and van der Veer 1993). But here, the important point is that Dumont attempted to note "certain overlaps" between the religious notion of purity and the mundane or atheist notion of hygiene. He may have borrowed this sense of overlapping from Srinivas, who argued from the other direction that overlaps suggested a pertinent distinction. Much earlier, in *Religion and Society among the Coorgs of South India* (1952), Srinivas argued that a distinction between cleanliness and purity was evident in native thought:

> It is necessary to stress that ritual purity is fundamentally different from cleanliness, though they overlap frequently. A simple association of ritual purity with cleanliness and ritual impurity with dirtiness would be a neat arrangement but it would falsify the facts. One comes across ritually pure robes which are very dirty, and snow-white clothes which are ritually impure. (1952:105)

Apart from Dumont's implicit orientalizing mission, a mission in which he was intent to show that Hindu thinking differed significantly from Western thinking, his reference to overlaps suggested that mundane reality, at some level, is encompassed by reverence for the sacred. Therefore, purity also exerts its effects at a mundane level by coding everyday events, many a part of sustenance and production. Taking his definitions from the *śāstrik* and *dharmik* (sacred and religious) literature, Dumont classified purity into three types, bearing on the family, objects of everyday use, and the body. About the latter, he wrote:

> For the body, the main thing is the morning attention to hygiene culminating in the daily bath. According to Manu, there are twelve secretions or impurities, including excrement, saliva, and the lowly fate reserved for the left hand. Objects are distinguished by the greater or lesser ease of their purification (a bronze vessel is merely cleaned, an earthenware one replaced) and their relative richness; silk is purer than cotton, gold than silver, than bronze, than copper. But above all one realizes that objects are not polluted simply by contact, but by the use to which they are put, by a sort of participation by the object, in being used, with the person. (1980:49–50)

This was Dumont's entry into the larger argument that sacred purity and impurity constitute the ideological system for social hierarchy. This means that to think in commensurate terms the Western reader or observer must shed materialist ideology and consider purity as the "umbrella" or ideological ideal for social order. In the passage that follows the first quote I provided, he extended the purity-impurity axis to account for social structure.

> Apart from the immediate physical aspect (cleanliness, hygiene) the etiquette of purity corresponds in one way to what we call culture or civilization, the less punctilious castes being regarded as boorish by the more fastidious. In relation to the social organization, those who are pure are in one way the equivalent of what we call 'decent' or 'well-born' people. In relation to nature we have indicated in passing how impurity marks the irruption of the biological into social life. Hence we find here a functional equivalent of that rift between man and nature which is so strongly in evidence in our own society and which seems to be unknown to, or even rejected by, Indian thought. Finally, the notion does not correspond solely to the prestigious, the beneficial, the auspicious (even though there are not only nuances but curious reversals): it is clear that in the general scale of values it tends to occupy a region which in our society derives directly from good and evil, but it introduces a relative rather than an absolute distinction: this offers an insight into the Hindu ethical universe. (1980:60–61)

For Dumont, Hindus use purity as an ideological mantle for separating groups of people on the basis of heredity and occupation. Yet, when making this point, he played on a distinction between elite and popular culture, echoing themes in early-twentieth-century European writing.[5] Carrying out his orientalizing mission, he argued that this ideological meaning also involves what a Westerner would call classifications of good and evil. Thus purity is at once an ideology, a social structure, and a transcendent power. The basic idea of the encompassing essence of the purity principle appears again and again in the works of other social scientists.

Pollution as Symbolic Marginality

In *Purity and Danger* (1966), Mary Douglas argued that rules of uncleanness must be considered in the context of a more elaborate conceptual system. For Dumont, hygiene was encompassed by the ideology of sacred purity. But Douglas also rejected a medical materialist view and considered uncleanness or "matter out of place" in terms of the range of powers and dangers symbolically constructed in a cultural universe. Douglas's goal, while less specific to South Asian conditions than Dumont's was, aimed to outline the human principles involved in the symbolic ordering of the universe. Order is created, she put it, by exaggerating differences between within and without, male and female, and other conceptual pairs. She argued that the recognition of ambiguity or anomaly—what cannot conform to culturally acceptable categories—causes anxiety and leads people to suppress or avoid the ambiguous or anomalous and to affirm the boundaries of accepted categories.

One of the problems is that Douglas used the terms *uncleanness* and *pollution* interchangeably. For anthropologists speaking in the English language, this made the cultural constructs one and the same in symbolic anthropology. Like Dumont, she also considered holiness and purity within the same category to avoid the tendency to separate holiness and hygiene into separate domains. Her passage on Hinduism addresses this point about interchanging terminology.

> In Hinduism, for example, the idea that the unclean and the holy could both belong in a single broader linguistic category is ludicrous. But the Hindu ideas of pollution suggest another approach to the question. Holiness and unholiness after all need not always be absolute opposites. They can be relative categories. What is clean in

> relation to one thing may be unclean in relation to another and vice versa. The idiom of pollution lends itself to a complex algebra which takes into account the variables in each context. (1966:9)

In this passage, Douglas does a very confusing thing, especially in light of points about overlaps made by Srinivas and Dumont. Douglas argued that it is ludicrous to put the unclean and the holy into the same linguistic category. By doing so she indirectly refuted Srinavas's idea of overlaps. Arguing this way, she opposed Frazer and other evolutionists who had advocated the view that primitive thought (which included Hindu thought) could not make a distinction between holiness and cleanliness. The evolutionists' view implied that primitives were steeped in a kind of naturalized affinity with the unfastidious. In *her* particular manner of orientalizing, Douglas collapsed uncleanness, unholiness, and pollution into a single cultural category and stood it up in relative opposition to cleanness and holiness. By doing so, she rather unwittingly reproduced the anthropological tendency to link religious impurity with pollution and to use the two terms interchangeably.

While interchanging the terms *uncleanness* and *pollution*, Douglas aimed to demonstrate that these cultures conflate sacred and secular domains, domains that, in Western culture, are held apart. At the end of her second chapter, she wrote:

> To conclude, if uncleanness is matter out of place, we must approach it through order. Uncleanness or dirt is that which must not be included if a pattern is to be maintained. To recognise this is the first step towards insight into pollution. It involves us in no clear-cut distinction between sacred and secular. The same principle applies throughout. Furthermore, it involves no special distinction between primitives and moderns: we are all subject to the same rules. (1966:40)

At many points in the latter part of the text, she substitutes uncleanness for pollution, showing that dirt, like social pollution, is matter out of place, on the margin, losing its identity. At the end of the book, she hints at the possibility of distinctions between dirt, uncleanness, and ritual pollution, but does not follow through with specific cultural definitions. As Argryous (1997:162) notes, Douglas assumed that dirt was a universal, negative value. She did this to work toward a primary argument: dirt fits within the category of the marginal, the dangerous, and the potentially creative. This leaves the reader with a model of symbolic ordering, but with no clear message about the meanings of dirt, ritual pollution, and uncleanness in non-Western cultures.

Auspiciousness against Purity

A primary criticism of Dumont's work is that he focused exclusively on caste and status and excluded analyses of other important cultural categories and their operations in Hindu life (see Fuller 1979; Carman and Marglin 1985). Nonetheless, many social scientists have adopted Dumont's definitions of purity and impurity. After extensive critiques of *Homo Hierarchicus* appeared, Dumont's definitions continued to underlie arguments about the distinctiveness of the Hindu worldview especially when lined up against the rather monotoned Western view. These definitions also appeared in arguments about the encompassing or comprehensive nature of signification practices and symbolic meanings. Meanwhile, studies of Hindu caste and ritual continued to use the English terms *impurity* and *pollution* interchangeably to code status and variables such as sin and evil, moral considerations, and relations between gods and humans (see Das 1982; Fuller 1979:473; Madan 1985; Marglin 1985a; O'Flaherty 1976). In some accounts, the term *ritual impurity* was dropped altogether in favor of *ritual pollution*. For example, Fuller wrote:

> Ritual pollution derives from many sources. For example, virtually all bodily emissions and waste matter are sources of pollution (saliva, semen, menstrual blood, feces, urine, hair, and nail clippings in particular). Organic life, in other words, is the most immediate source of pollution, which is mainly controlled by a daily bath and other ablutions as required. But pollution is also controlled by allocating to specific, low-ranking castes duties like barbering, laundering, and removing nightsoil, so that the purity of the Brahmans and high-caste people is preserved by others who perform polluting tasks for them. (1992:15)

To expand scholarly debates beyond the focus on purity and impurity, others began to describe how the Hindu notions of auspiciousness *(śubha)* and inauspiciousness *(aśubha)* work through the religious worldview. Madan (1985) discussed the ways in which Kashmiri Brāhmaṇs used the terms for auspiciousness *(śubha)* and purity *(śuddha)*, arguing that purity could be more easily objectified than auspiciousness could. Purity is altered by human effort, while auspiciousness results from a series of intersections and conjunctions that individuals have little control over. About purity he wrote:

> The word *śuddha*, in contrast to *śubha*, is not generally used in everyday speech to refer to events. The connotation of this word is conveyed by invoking images of fullness or completeness in the specific sense of perfection. It thus refers to the most desired state of being. *Śuddha* and its opposite *aśuddha* are attributes of animate beings, inanimate objects and places with which a human being comes into contact in the course of everyday life. For example, a prepubescent unmarried girl *(kanyā)*, water from a holy river, unboiled milk, ghee and a temple are *śuddha*. On the other hand, contact with certain kinds of human beings (low caste Hindus or non-Hindus), animals (dogs), objects (goods made of leather), foods (beef or food cooked in impure utensils), substances (discharges from a human body), places (cremation ground), etc. causes Brahmans and other upper caste Hindus to become polluted. The notion of perfection in the sense of freedom from error or fault is extended to certain actions also as is exemplified by such expressions as *śuddhavicāra* (pure thoughts), *Suddhauccāraṇa* (correct pronunciation which is highly valued in the recitation of sacred texts), and *śuddhasvāra* (normal or natural notes in music). Human thoughts, words and musical notes are thus treated (evaluated) as objects are. (1985: 17)

In this passage, Madan tied his definition of purity closely to ritual status and to the status of objects and persons. He did not mention connections with hygiene, sanitation, or medical materialism but included them in the broader notion of fullness or completeness that purity stands for. Like Dumont and Douglas, Madan defined his categories as relational entities, suggesting that auspiciousness *(śubha)* and purity *(śuddha)* were always constructed in relation to the *pātra* or actor.

In these scholarly debates, we also find that a subtle shift in the notions of sacred and secular was under way. But this shift was not being scrutinized in the same way that the purity-pollution axis was. In an eloquent address to the Asian Studies Association in 1987, Madan proposed a more contemporary notion of encompassment and used it to suggest a different configuration of the sacred and the secular. He was moving beyond Dumont's notion of encompassment by revising it to apply not to caste relations but to relations between religion and the state and institutional and organizational movements (see chap. 10). Nonetheless, the potential for readers to link Dumont's sense of the sacred and secular to the idea of encompassment remained, and this has threatened to confuse Madan's meaning. In an earlier work, Marglin (1977) charged that Dumont had made an artificial separation between status and power. Her treatise helps us to consider Madan's meaningful references to the institutional tensions between secular ideology and what he referred to as "the religious" in that 1987 lecture.

While Madan's reformulated meanings of sacred and secular are appreciated, it is still problematic that he and others have conflated the term *pollution* with *ritual impurity*. This has only clouded the Hindu distinction between ritual impurity and material dirtiness. Understanding *pollution* becomes even more problematic when scholars attempt to differentiate, ethnographically, the meanings of environmental pollution and ritual impurity when they occur in the same context.

Environmental Pollution

Recent studies of Hinduism have highlighted the ecological functions and meanings of religious beliefs and rituals to show that, historically, Hinduism has embodied a concern for nature (see Chapple 1998; Chapple and Tucker 2000; Cremo and Goswami 1995; Narayanan 2001; Nelson 1998). Curiously, many of these studies have continued to stress the relation between sacredness and ecological wholeness that Madan elaborated, without looking at how ecological marginality (waste, dirtiness, and so forth) might be coded and treated in religious discourses. Other studies have translated Hindu categories and other cultural classifications of ecology into Western environmental frames to suggest that a degree of compatibility exists. The applied agenda behind this is to promote a global dialogue and create a cross-cultural environmental ethics, a syncretic doctrine for all of humanity in this age of environmental crises.[6] Differing from that of Dumont and Douglas, this round of scholarship seeks commonality rather than distinctiveness.

While I agree that religion plays an important role in shaping perceptions and understandings of the environment, I argue that the problem must be approached from the other direction. Rather than begin with a footing in naturalist notions such as the "environment," a more thorough deconstruction of the religious frame must lead the discussion. Then considerations of dirtiness, uncleanness, waste, and environmental pollution can proceed. This is why I begin with a focus on Hindu discourses to flesh out the variants of this religious frame, then widen my scope to interactions with official, scientific, and activist discourses. While I argue that a single environmental movement against river pollution is not emergent, I do suggest that certain overlaps, in Srinivas's sense, do occur in the discourses of religious leaders, government officials, politicians, environmentalists, and others. Two separate ideas can, under certain conditions, mean the same thing. Likewise, a single term can mean different things to different speakers and listeners. However, this does not require that the languages used, the frameworks of knowledge referred to, and the traditions of communication practiced be all the same. If this were the case, then the collective action of an "environmental movement" *might* have already begun.

CHAPTER 3

The Power of Gaṅgā

THE GAṄGĀ RIVER AND ITS TRIBUTARIES drain more than one million square kilometers of China, Nepal, India, and Bangladesh. In India, the Gaṅgā basin, which includes the Yamunā subbasin, covers over 861,000 square kilometers, or one-fourth of India's geographical area (Das Gupta 1984:5). The Gaṅgā river begins at the confluence of the Bhagirathi and the Alaknanda, located within the Himalayan foothills of northern India. The Bhagirathi flows from the Gangotri glacier at Gomukh and the latter from a glacier near Alkapuri. Flowing across the great alluvial Indo-Gangetic plains, the Gaṅgā is bordered by the Himalayas to the north and the Vindhya-Satpura ranges to the south.

The main stem that carries the name Gaṅgā is entirely within India and this channel and its water are most sacred to Hindus. Along the river's traverse, large tributaries enter the Gaṅgā and significantly increase its flow and change its character. The Gaṅgā is joined by the Ram Gaṅgā, Yamunā, Ghaghara, Gomti, Gandak, and Kosi tributaries. However, the major tributaries, the Ram Gaṅgā, Yamunā, and Ghaghara, are the only Himalayan rivers that have significant base and flood flows.[1] Water also comes from snowmelt and from monsoon season rainfall. The rivers of the Gaṅgā basin carry one of the largest sediment loads in the world. Today sediment loads in the Gaṅgā are higher than in the past due to the complete deforestation of the Gangetic Plains and the ongoing deforestation of the Himalayan foothills.

The Upper Ganga Plain is one of the most highly irrigated regions of the country. The Upper Ganga Canal irrigates two and a half million acres of agricultural land in Saharanpur, Muzaffarnagar, Meerut, and Bulandshahr districts. Moreover, these waterways stretch across one and a half million acres of the central and lower Doab (Shukla and Vandana 1995:8). The Gaṅgā is dewatered for these irrigation purposes at several sites. The most upstream site is located at Haridwar, where the plains meet the foothills of the Himalayas. At Haridwar, a significant portion of the main stream is diverted into the Upper Ganga Canal. This is an irrigation channel that feeds the alluvial tract lying between the Gaṅgā and Yamunā Rivers. The long canal system through the Doab has two names: the upstream part is referred to as the Upper Ganga Canal, while the downstream section, starting at Aligarh, is the Lower Ganga Canal. At Kanpur, the irrigation return flow reenters the parent stream. Scientists argue that the reach of the river from Kannauj to Allahabad is particularly vulnerable to human-induced pollution because significant dewatering has occurred upstream. Beyond this reach, the Gaṅgā gains stream flow from the Yamunā River at Allahabad and from the Ghaghra, Kosi, and Gandak Rivers farther downstream (Shukla and Vandana 1995:234).

The level of pollution in the river depends upon the concentration of pollutants and the discharge of the river. Both concentration and discharge are affected by hydrological, geomorphologic, topographic, and cultural factors. Today over 45 million people reside in the Gaṅgā basin. From the Himalayas to the Bay of Bengal, the Gaṅgā passes alongside 29 cities with populations above 100,000 (known as Class I cities), 23 cities with populations between 50,000 and 100,000 (Class II cities), and 48 towns with populations below 50,000 (Central Pollution Control Board 1995a). The heaviest concentration of urban centers lies in the Doab. The Gaṅgā provides the greatest source of drinking water for the cities, towns, and villages along the riverbank and is also the greatest source of water for domestic, municipal, and industrial requirements. Shukla and Vandana (1995:49) estimate that the demand for water from the river currently stands at more than 62 billion cubic meters per day.

These human settlements also use the waterway to dump human, animal, and industrial wastes. The Central Pollution Control Board reports that the main sources of pollution that find their way to the river are urban liquid waste (sewage/sullage), industrial liquid waste, surface runoff from solid waste landfills and dump sites, and solids and liquids from practices such as bathing of cattle and immersing dead bodies in the river. The Board (1995a:1) reports that three-fourths of

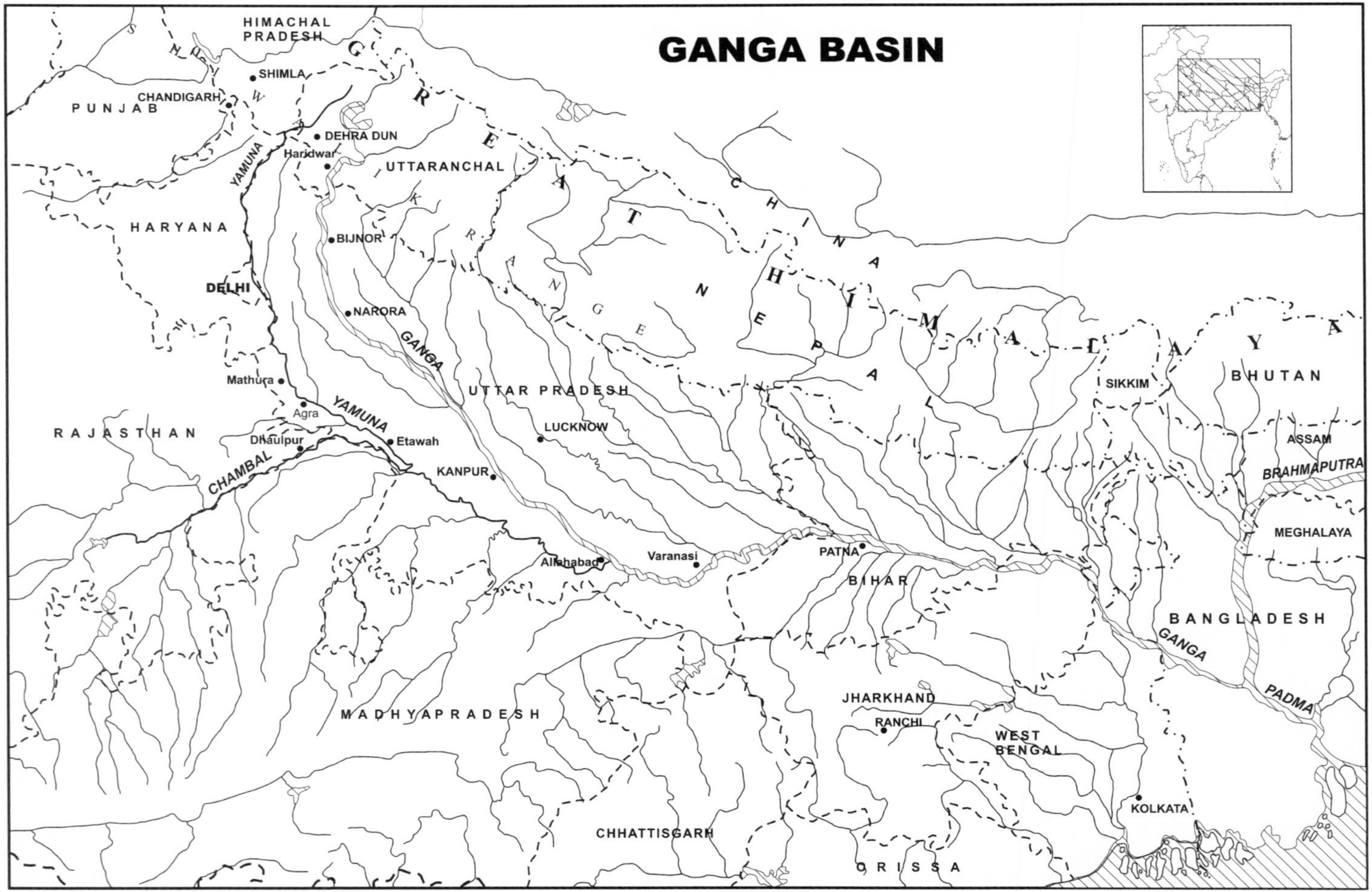

Map 1. The Gaṅgā basin. (Map source: World Wide Fund for Nature—India and Paani Morcha, modified by Cyrus Dawsey.)

the pollution of the river comes from the discharge of untreated municipal sewage, and 88 percent of this is created in Class I cities. However, without more specific data on discharge amounts, the Central Pollution Control Board is limited to generating estimates, tallied from the approximations of wastewater flows in each of the cities through which the river passes. Part 4 discusses some of these estimates and the groups who construct them.

The Upper Gaṅgā Plain running through the state of Uttar Pradesh is the most industrialized part of the river basin. This region of Uttar Pradesh is home to sugar factories, leather tanneries, textile industries of cotton, wool, jute, and silk, food-processing industries related with rice, dal, and edible oils, paper and pulp industries, heavy chemical factories, and fertilizer and rubber manufacturing units. Industrial wastewater is discharged by a number of these industries. Four major thermal power plants also depend upon water from the Gaṅgā. In 1995, the Central Pollution Control Board listed 191 grossly polluting industries in the state of Uttar Pradesh, 6 in the state of Bihar, and 67 in the state of West Bengal. At the time the report was made, these industries were discharging toxic effluents that carried biological oxygen demand (BOD) concentrations of more than 100 milligrams per liter. According to the board's reports, each unit's discharge added up to more than one million liters of wastewater per day. Researchers have found heavy metals such as cadmium, zinc, nickel, lead, chromium, and copper concentrated in the river water and the sediments. In 1995, the Central Pollution Control Board (1995b:14–15) reported high levels of zinc and nickel in the sediment of the river Yamunā, a river that flows past the capital city of Delhi and into the Gaṅgā at Allahabad. Researchers from Nehru University in Delhi have also reported PCB concentrations as high as 782 parts per billion and have detected the presence of DDT, hexachloro hydrocarbons, and cyclodiene in the wastewater entering the Yamunā River.[2]

Algae form the base of the food web that links the Gaṅgā's ecosystem with biological communities. Approximately 577 algal species are known to be present, playing two roles in the ecosystem. First, algae play a vital role by consuming carbon dioxide and producing oxygen during photosynthetic activity. The released oxygen dissolves in the water and increases the oxygen level needed for other plant life and human uses. Algae may also absorb some dissolved chemical substances that are toxic pollutants and use them as nutrients for their growth. Substances like calcium, magnesium, nitrate, nitrite, potassium, and iron are consumed as nutrients by algae also during their growth. However, algae also act as a source of pollution because they can consume oxygen present in the water, have toxic effects, spread allergenic diseases, and produce odors and colors (Shukla and Vandana 1995:226). Many scientists are now looking into ways that microbiological features can be used to help regenerate the ecosystem and assist in pollution abatement.

Purificatory Power

For at least two and a half millennia, the Gaṅgā basin has served as the seat of human civilization. Buddhist records explain that in the sixth century B.C. the valley was a large jungle that attracted agriculturalists looking to colonize the resource base of hunter-gatherer populations (Gadgil and Guha 1992:78). By the fourth century B.C., Pataliputra (now near Patna, the capital of the state of Bihar) was one of ten ancient capital cities of India. Supported by heavy trade along the river, Pataliputra dominated the Gaṅgā basin. Later, under the leadership of Ashoka, the Mauryas unified the valley from Haridwar to Tamralipti and kept their capital at Pataliputra. Book two of Kautalya's *Arthasastra* (ca. 321–300 B.C.), written during the Magadha Empire of Candragupta Maurya, describes vigorous economic activity along the Gaṅgā. Remaining at the epicenter of the great dynasties that rose and fell through the mid–sixth century A.D., the Gaṅgā served as the main waterway for long-distance trade and transportation.

The Gaṅgā basin is also the cradle of Hindu and Buddhist pilgrimage culture. Some of the most important centers of spiritual learning and healing have developed along her banks and those of her tributaries. At the headwaters of the Gaṅgā in the Himalayas, sacred shrines at Tapavan, Gomukh, Bhojbasa, and Gangotri mark the sources of her power. The shrines of Kedarnath and Badrinath also celebrate their position in the upper reaches of the watershed. Farther downstream in the Himalayas are Uttarkashi and Rishikesh, and along the plains lie Haridwar, Allahabad (Prayāg), Banaras, Nadia, and Kalighat. Along the Yamunā River, we find the sacred complexes of Mathura and Vrindavan and, inland within the Gaṅgā basin, the Hindu center of Vindyachal and the Buddhist sites of Gaya, Rajgir, and Nalanda. At these pilgrimage centers and at countless other smaller sacred spots along the great traverse, pilgrims worship Gaṅgā and carry away her water for use in worship and purification rituals.

Hydraulic power is central to the Hindu understanding of Gaṅgā's sacrality. Her hydraulic power is generally contrasted with the flushing power of other rivers on the subcontinent. The Gaṅgā, unlike the Yamunā that flows into it, is a fast-flowing, shallow river. The Yamunā is a slower, deeper river and flows closer to the plains than the Gaṅgā does before their confluence at Allahabad. Residents of the Gaṅgā-Yamunā basin distinguish between the two rivers by the color and texture of their flows. The Gaṅgā is grayish and silty while the Yamunā is green and cold. Residents point to the visible differences between the two at the point where they converge at Allahabad or Prayāg. Visible differences give Hindus evidence for the argument that the two rivers are distinctly different goddesses and remain separate entities even after their confluence. According to the Hindu view, the sacred is not detached from the material *(bhautik)* realm of ecology and the built environment, but provides a context for understanding the truth of the physical world. Interpretations of sacred space describe time-space conjunctions between divine power and the physical world. Hindu mythologies and sacred texts also weave together understandings of sacred and natural ecology, to portray the intersection of the sacred and the material with human culture.

The meanings of the Gaṅgā embedded in the sacred texts of Hinduism provide a symbolic context for understanding the perspectives of many of the informants I introduce in this book. Connections between these textual interpretations and the discourses and ritual traditions of people gathering along the riverbank fill out the cultural milieu. Religious leaders and pilgrim service providers describe Gaṅgā as a goddess who absolves worldly impurities and rejuvenates the cosmos with her purificatory power. She is also a mother who cleans up human sin and mess with loving forgiveness. Hindus show their respect to her in oil lamp rituals *(āratī)* that they perform on the riverbank and in temple worship *(pūjā)*. Most important, devotees seek spiritual purification by doing ritual ablutions *(snān)* in the river.

Gaṅgā intersects with physical space at many locations. The more auspicious conjunctions are marked by the built environments of urban-sacred complexes. Banaras, called Varanasi in official discourse, is an important pilgrimage-urban-sacred complex bordering the Gaṅgā in the eastern part of the state of Uttar Pradesh. For Hindus, it is the center of Śivā's universe, as well as the beginning and end point of human civilization. Hindu mythology has it that Kashi, the ancient name for Banaras, contains the whole world and everything on earth that is powerful and auspicious. It contains the Mahāśmāśāna, the great cremation ground that survives the destruction of the cosmos brought about by Śivā's ascetic power (see Parry 1980:89; 1981:339).

Most residents of Banaras, whether they work in government service or private business, emphasize Gaṅgā's sacred purity and call upon her divine power in worship rituals. They invoke Gaṅgā's symbolic history when reciting eulogies to her recorded in sacred texts such as the *Rāmāyaṇa*, *Mahābhārata*, the *Purāṇas*, and *Māhātmyas*. Gaṅgā is also inscribed in temple sculpture and art as a purifier, mother, sustainer, and daughter or cowife of Siva (see figs. 1 and 2).[3] The *Ṛg Veda* describes Gaṅgā's character as a life force and goddess.[4] In the *Mahābhārata*, Gaṅgā takes anthropomorphic forms as daughter of Bhāgīratha, mother of Bhismā, and wife of Samtanu.[5] Puranic inscriptions cite the powers she can offer devotees and outline how a bath in Gaṅgā cures ailments, makes impure people pure, and leads to *mokṣa* (final liberation). These prescriptions are linked to the ritual practices of ablution through behavioral gestures and verbal admissions of sentiment (Dimmit and van Buitenen 1978).

Many stories I heard of Gaṅgā's descent from heaven related the main events of the Gaṅgā-avataraṇa, a chapter in the *Rāmāyaṇa*. They described how a devotee named Bhāgīratha called Gaṅgā down to earth to purify the ashes of King Sagara's 60,000 sons. King Sagara once reigned across northern India and would periodically send a horse out across his territory to reaffirm his power. If the horse were to roam free and return safely, the king's sovereignty would be established. To capture the horse was to challenge the king's rule and lay claim to the territory. On one test, the king's horse was captured by a *rishī* (renouncer and monk) who was meditating on the banks of the Bay of Bengal. King Sagara's sons were sent out to recapture the horse, but never returned. The rishī, insulted by their audacity, had burned them to the ground with his rage.

According to a more unique telling of this story, Sagara had originally wanted daughters rather than sons, as blessings from God. Sagara

Fig. 1. Contemporary outdoor exhibit of Gaṇgā's descent from heaven

Fig. 2. Sculpture of Mother Gaṅgā (eighth century)

hoped to gain daughters and give them away in a meritorious practice known as *kanyādān*.

> There was once a jackal in a forest who gave his daughter in marriage to another jackal in another forest. After some time this jackal went to visit his daughter. The forest was a big one and he thought the forest was their whole house. So he didn't want to eat or drink anything in that forest. He was adhering to the principle very strictly [that food and gifts were not to be taken from the other family in exchange for the daughter]. When he was coming back to his forest, he died on the way. When he died he became a human being and he remembered his past life, that he was a jackal and that by giving one daughter in marriage and observing the rules he became a King, a human being. Suppose, he thought, I do such sacrifice of *kanyādān* with 60,000 . . . and he calculated: by doing one kanyādān how much merit, by doing two how much? By doing 60,000, he could become the King of the Gods! To have 60,000 daughters, however, was not humanly possible so he thought, let me pray and get divine grace. Ultimately he found that God was pleased with him and told him he could have whatever he wanted. But by that time the King of the Gods had become jealous because he thought this man might take his position. So he went to the Goddess Saraswati, Goddess of learning and wisdom, and told her to go and sit in the mortal's mind. The God told her: he is going to ask for *putrī* (daughter). Tell him instead of *putrī* to ask for *putrā* (son). So the King was doing penance and the God was pleased with him and asked, What do you want? The King replied, I want 60,000 *putrā*. His intention was for *putrī!* But he asked for sons and that is how Sagara got 60,000 sons. Instead of giving them in marriage and getting merits he trained them in warfare.

After this, the story proceeds according to the more common rendition. In all versions, King Sagara's sons are finally liberated by the grace of Gaṅgā. After many generations, Bhāgīratha, a descendant of this great king, achieves sufficient power through his devotional penance, enough to impress on Lord Brahmā that his wish be granted. Bhāgīratha asks that Gaṅgā descend to earth and purify the burning coals of his ancestors. Brahmā agrees and pours Gaṅgā out of his jug and onto the locks of Lord Śivā. From there, she meanders down the peaks of the western Himalayas and follows Bhāgīratha across the plains of northern India to the site where the burning coals lay. At that spot, she purifies them by her touch and then flows into the Bay of Bengal.

There is another descent story associated with this one. The *Bhāgavata Purāṇa*, a treatise of the medieval period, relates how Viṣṇu in his incarnation as a dwarf asked for three strides from Bāli, the Lord of the Universe. Thinking the dwarf's stride would be limited, Bāli granted the wish. But then the dwarf (who was really Viṣṇu) grew exponentially and extended three legs across this world, the heavens, and the netherworld in the strides allotted by Bāli. The toe of one foot scratched the highest tip of the cosmic egg encapsulating this world, and out of this crack Gaṅgā flowed and washed down Viṣṇu's foot. The verse relates how, in the descent of rivers from the cosmic realm, the sins of the world are washed away.

> In the sacrifice performed by Bāli, the Lord (Viṣṇu) himself appeared as Trivikrama. Standing there on the ground, he wanted to measure three feet of land donated to him by Bāli. He covered the whole of the earth by his right foot. He raised the left foot to measure the heavenly regions by his foot-step, but the upper crust of the shell of the cosmic egg got cracked by the nail of the big toe of his left foot. Through that opening, rushed in the stream of waters, covering externally the cosmic egg. While washing the lotus-like feet of the Lord, she (the water of the stream) became reddish by the pollen-like dust (on the Lord's feet). She washed away the dirt, in the form of the sins of the whole of the world, by her touch, and yet, herself remained pure (unpolluted by sins). (*BhP* Skandha 5 17.1; trans. Tagare 1976:716)

After washing over Viṣṇu's foot, she descended onto the head of Lord Śivā. He contained her powerful flow with his locks of hair. In this way, Gaṅgā's descent is intricately related to Brahmā, Viṣṇu, and Śivā; Gaṅgā herself is called *tripathagā* (the union of three paths or aspects). The Śaṅkarācārya of Jagannath Puri, a highly respected religious leader, explained the meaning of tripathagā in a taped interview.

> We worship Ganga's birth from Brahma's *kamaṇḍal* (jug). Gaṅgā is *tripathagā* (the union of three paths or elements); she holds influence in the three worlds of heaven, earth, and hell. From the foot of Lord Viṣṇu, she became manifested, therefore she is related to Bhagwān Viṣṇu. Śrī Brahmā brought her down on the foot of Viṣṇu therefore Brahmā made a contribution to the manifestation of Gaṅgā. Bhagwān Śivā contained Gaṅgā's flow with his locks and so she is also connected with Śivā. The creator of this world is Brahmā, the Sustainer of the earth is Śrī Viṣṇu, and the destroyer of the universe is Śivā and Rudra. There is one great principle in the form of these three Gods, and this is expressed in the powerful dance of Gaṅgā's descent *(acit līlā śakti ke yog se avtārit hua hain)*. In this view, the self-forms of Brahmā, Śivā, and Viṣṇu are the avatārs of the Great Soul and Great Brahman. By the grace of all three, Gaṅgā was manifested.

Gaṅgā's immanent form as water *(jala)* is a central element of ritual practices for Hindus. Snān in the early morning hours is an essential component of daily ritual for residents and pilgrims in sacred cities along the river. Hindus also immerse the ashes or bones of the cremated to ensure their safe journey to the realm of the ancestors. Uses of Gaṅgājal for pūjā (offerings to deities) affirm the eternal essence of the river. Devotees also carry Gaṅgājal in jugs to temples where they perform *jalābhiṣek* (pouring of Gaṅgājal over a Śivā *liṅga*, the iconic, phallic form of Śivā) and other worship rituals. These ritual uses of Gaṅgājal are distinguished in residents' minds from the uses of her made through the municipal "filtered water" system.

In sacred places, pilgrims and residents also revere and seek blessings from Gaṅgā by undertaking the ritual of *āratī*, a ceremonial offering of fire. Devotees, while standing on the riverbank, wave an oil lamp and other sacred objects in front of Gaṅgā (see fig. 3). The sounds of bells, gongs, drums, and conch shells play a prominent role. A devotee gains power by chanting *mantras*, formulaic phrases understood as verbal codifications of sacred texts. Singing also accompanies the last rites of the āratī ceremony. On Pañcgaṅgā ghāṭ in Banaras, Brāhmaṇ priests and pilgrims perform āratī every day at sunrise and sunset and sing the following *stuti* (hymn):

> Salutation, darsan, puja removes sin *(pāp)*

King Bhagirath did this for the well-being of humanity *(lokkalyān)*
Giver of devotion, liberation, peace, and freedom
King Bhagirath did this for the well-being of humanity
Auspicious deeds and service done
King Bhagirath did this for the well-being of humanity
Remover of distress, danger, and provider of wealth
King Bhagirath did this for the well-being of humanity
Knowledge of Brahman, *yajña* [ritual sacrifice], complete understanding
King Bhagirath did this for the well-being of humanity
Karma, knowledge, and devotion *(bhaktī)*
King Bhagirath did this for the well-being of humanity
Ramanandi, free from all worldly attachments,
was made the teacher of Vaishnav
Sarvodaya [well-being for all], Śivā in this world, auspicious, all being, transcendent

Devotees celebrate Gaṅgā's purificatory power and abiding grace in the festivals of Gaṅgā Daśaharā and Gaṅgā Saptamī. However, the performance of ritual and the celebrations of festivals do not have any effect on her purificatory power. Rituals are not powerful enough to purify Gaṅgā in a kind of feedback process because devotees see Gaṅgā's purity as part of a more holistic process of cosmic order and balance. This is an order in which humans strive to live harmoniously (see also Fuller 1979:460; 1992:76). Although some informants did mention that saints are able to generate the power to purify Gaṅgā in return (see also Vatsyayan 1992), most admitted that the average individual has no such power. Still, pleased by their efforts, Gaṅgā blesses the faithful, purifies their minds and souls, and grants a devotee's wishes if requests are made with pure faith.

When asked how the sacred texts *(śāstras)* guide their relationships with Gaṅgā, many Hindus pointed to rituals of snān and pūjā and the embeddedness of Gaṅgājala in everyday life. Snān works to clean physical dirt through the presence of or association with spiritual power (Dumont and Pocock 1959:28), thus reproducing at one level an interlocking relationship between ritual, purity, and physical cleanness. As one merchant working on Daśāśvamedha ghat in Banaras put it, "People who do bathing, meditation, and ritual worship (*snān*, *dhyān*, and

Fig. 3. Priest performing āratī on Pañcgaṅgā ghāṭ

pūjā) of gods and goddesses, this is understood as our history and knowledge." Dumont and Pocock pointed out that one must be pure to approach gods and goddesses (1959:30; see also Kane 1973:563–64). Although humans cannot reach the purity of the divine, there is some expectation that purity is a condition for that contact to be beneficial.

There are two views in current popular discourse about whether Gaṅgā can purify an impure person. One view argues that an impure person cannot become pure simply by bathing in a sacred place; rather, individuals must engage in the more holistic purifying process of committing the soul to Śivā. The other states that by reciting the name of Gaṅgā, one gains mastery over sin, by taking *darsán* (auspicious sight) one achieves well-being, and by performing snān one purifies seventy generations. The *Gaṅgā Stuti*, a eulogy sung to Gaṅgā, communicates these themes. In Banaras, at the time of Gaṅgā Mahāratī (great āratī) performed on the day of Kārtik Purnimā (the full moon day concluding the month of Kārtik), pilgrim priests sing this song:

From the place where the lotus foot of the Lord, where Bhagirath did great *tapasyā* (ritual austerities), Gaṅgā poured out from Brahmā's jug into the locks of Śiv-Śaṅkar.
She then descended to earth on a mountain of countless sins.
Tulsī Dās says open your two eyes and see how naturally she flows as a stream of nectar.
Those who take her name in memory will get *mukti* (liberation), those who do *praṇām* (prayer) will arrive at God's place.
Those who come to the banks of Gaṅgā will find heaven, those who see the waves of emotion will get *mokṣa* (the transcendent).
O Taraṅginī (Gaṅgā), your nature is this, what God has given to you.
O Bhāgīrathi (Gaṅgā), I am full of sin and dirtiness. I believe you will give me *mukti* and a place at your feet.

This stuti communicates themes similar to those developed in the *Gaṅgā Leharī*, a poem written by Jagannath in the nineteenth century (Chinmayananda 1997). For mortals, Gaṅgā is a vehicle to *mokṣa* and is the transcendent herself.

In *Water and Womanhood* (1995), Feldhaus argues that when interacting with rivers, many residents of Maharastra stress a river's female attributes over her purificatory powers. Banaras residents also conceive of the river in feminine terms. Many Daśāśvamedha residents and pilgrims claim that Gaṅgā, like a good mother, cleans up after the messes her children make and forgives them lovingly. In this way, she cleans up other kinds of dirtiness people bring to her, excusing them with maternal kindness. However, men generally do not accentuate the powerful aspects of motherhood generated through reproduction, focusing instead on her selfless cleaning and forgiveness. This attribution of motherly duties to Gaṅgā may be part of a male strategy (that has support in the sacred texts) to associate women's responsibilities with these tasks. It is also a reflection of what women are already doing: nourishing and cleaning up after children. In this way, Gaṅgā cleans up the messes of the whole of humanity, selflessly and lovingly. This understanding is one that I will return to in chapters four and ten, and again in the conclusion.

Science and Religion—A Convergence?

The two aspects of Gaṅgā's power—the material *(bhautik)* and the spiritual *(adhyātmik)*—intersect to form two complementary aspects in the Hindu worldview, even as they are associated with the distinctly different domains of science and religion in popular discourse. Spiritual aspects, many religious leaders claim, also reflect or embody scientific theory. And many point out (see Bandyopadhyay et al. 1985:159) that the religious heritage of Indian society provides rich ecological knowledge that includes rules for natural resource utilization.

Analyzing events in India's colonial and postcolonial periods, van der Veer (1994:162) has highlighted the constant interaction between, as he put it, "the empiricist insistence on the 'facts,' scientifically established, and orientalist and nationalist narratives." Although he focuses specifically on disputes over sacred spaces, the interplay he describes is also evident in the intersections between science and religion in discourses about the Gaṅgā. Van der Veer outlines how this interplay has occurred in violent struggles to appropriate sacred sites in Somanatha and Ayodhya. Such struggles have brought scientific and religious ideologies closer together as handmaidens in acts of violent destruction. All parties in these disputes have employed a mix of scientific logic and religious metaphor. In conflicts over Ayodhya, the community of scholars and professionals who argued for a secular-scientific understanding of the past had to, at some level, acknowledge the explanations of religious nationalists even while arguing against them. Scholars in the Department of Archaeology at Jawaharlal Nehru

University cautioned for restraint and discouraged religious leaders from avenging the past through the destruction of material remains. But others closer to government circles—B. Lal, for example—established a national archaeology project along Hindu nationalist lines. This project, "Archaeology of the Rāmāyaṇa Sites," claimed that Ayodhya was the birthplace *(janmabhūmī)* of Ram, king of the *Rāmāyaṇa*. Yet curiously enough, neither type of scientist could unearth or produce material evidence for their arguments. In the absence of material evidence, scientists did little more than make references to imagined pasts of their own, and in their weakness they were encompassed by religious narratives.

Van der Veer explains how empirical facts were used in this case to fill a lacuna in a well-known religious narrative. Scientists attempted to provide material evidence for the authenticity of a powerful religious narrative, the *Rāmāyaṇa*. While they hoped to encourage an excavatory interest in locating the past in material remains, in effect they promoted the opposite. Religious narratives about Rām's birthplace were strengthened by the fact that scientists were willing to follow the issue and enter the debate without material evidence. These narratives and the television serial the *Rāmāyaṇa* following later that year fed the fervor that destroyed Akbar's Mosque in 1992.

Assessments of the river Gaṅgā are just as interactive as conflicts over sacred space have been, though they are not tinged with the same level of violence. However, while interactions in this field are not violent, they are hardly apolitical. Like decision making in the development programs in Lesotho (Ferguson 1990), decisions about what it means to "save" an environmental resource are public appeals that carry out very private projects. In the process of appearing apolitical, these appeals and decisions contest use values and transform the very identity of the Gaṅgā.

As religious narratives capture a hegemonic position in discourses about sacred space, they are also the dominant narrative mode in public discourses about the river. Some Hindus living in cities in Uttar Pradesh invoke the scientific notion of water quality to support the master narrative of Gaṅgā's sacrality. This master narrative is embedded in the Gaṅgājal story and supported even when Hindus invoke the modernist notion of microbial impurity. Many people will argue that if kept in a jar or glass at home for years and years, Gaṅgājal will never "spoil" or "develop bacteria." It will never develop the kind of bacteria that breeds in mineral water or tap water. Extending the notion of microbes further in this trajectory, a student from Banaras-Hindu University once told me that an unidentified microbe found only in Gaṅgā was responsible for the river's incredible self-purifying capacity. Gaṅgā possessed a special microbe that was able to eat up material pollution at a fantastic rate. But the nature of this microbe was a spiritual issue for him: why did Gaṅgā, above all other rivers, possess it?

The master narrative of Gaṅgā's sacred purity is reproduced most convincingly in the ancient sacred cities bordering the river. In Banaras, Allahabad, and Haridwar, master narratives about Gaṅgā's purificatory powers are tenable as personal and family convictions because they are reproduced through occupations *(pesá)* and everyday ritual practices. Because these narratives tend to disregard surface oscillations in favor of the longue durée, proponents are able to argue that Gaṅgā's purificatory power can outlast a great deal of material assault. For them, the longue durée brings out the true essence of Gaṅgā's purificatory power. So religious narratives encompass material conditions by creating a place within their larger frameworks for the secular concern with water quality. In this place, science is used to reaffirm the power of Gaṅgā's purity and of Hindu *dharma* (order and duty) more generally. This encompassed, material place is described by the shifting bhautik realm, a world eclipsed by more powerful cycles of cosmic order.

On the scholastic side, we also find the intersection of religious codes and scientific assessments. In academic and government research and policy reports, scientists and government officials invoke religious meanings of the river to justify the utility of water quality studies and sewage treatment systems. While officials generally dismiss any relevant connection between ecology as it is known through science and the cosmic theory of Gaṅgā's origin and essence, officials do not cast off religious symbols as if they had no valence among the wider public. This is evident in reports published by government-appointed university research teams. In a report on the Gaṅgā basin prepared by the Central Board for the Prevention and Control of Water Pollution, one author began by quoting a Sanskrit *sloka* (verse) by Valmiki, then provided an English translation.

Pāpāpahāri duritāri tarangdhāri durpracāri girirājguhāvidāri
jhan kārkāri haripādrjoviddāri gānmyam punātu satant subhkāri vāri

Let the pure murmuring waters of the Gaṅgā, which removes the evil, destroys the sin, takes away the dust from the feet of (Lord) Harī (of Haridwar) and runs in waves to far distances, having pierced through the depths of the Himalaya, cleanse (us) ever—Valmiki[6]

Other reports followed, fitting data on the physicochemical and biological profile of the river into a larger cosmic framework. The section on hydrological characteristics began with a sloka that gave the scientific document a literary charm but left open the possibility of a transcendent power beyond knowledge of natural ecology.

Bhāgīrathi sukhdāyini mātasvah jalamahimā nigame khyātha
Nāham jāne tab mahimānam strāddi kripāmayi māmkshānam

Oh Mother Bhāgīrathi (Gaṅgā), the source of delight! Your wonderous waters are well known in the books of knowledge. I can not fathom your greatness. Deliver me from my ignorance, oh merciful one!—Śaṅkarācārya[7]

Following this theme, a report titled "Gaṅgā, the Most Self-Purifying River" explained that Gaṅgā has a mystical transcendent quality that befuddles even the most serious scientists. The scientist who authored the report called Gaṅgā the most "naturally purified" river. He wrote:

During a water quality survey of Ganga and Yamuna, several interesting phenomenon [*sic*] were observed which played significant role in the natural purification of these rivers. The rate of exocellular polymers, excreted during the endogenous growth phase of the various species of bacteria, vary considerably by reducing the BOD and hence turbidity. The interaction of exocellular polymers present in the rivers with the colloidal matter present in sewage results in a very rapid and significant reduction of the BOD. The data analysis has shown that Ganga has the highest BOD rate constant value and the reaeration rate constant value, both of which are higher by an order of magnitude than the thus far reported values [for any river].

> The coliform organisms are also found to reduce significantly in very short times. The findings have been reasoned to suggest that Ganga is the most naturally purified river. (Bhargava 1981a:233)

In these and other passages, scientists have argued that Gaṅgā carries a high bacterial load but is able, by aeration, to reduce these levels in a very short time (see table 1). Nath and Nath (1990:xvi) have also pointed out that "men of science" from overseas recognize the Gaṅgā's special "pathogen killing properties" and admit that an explanation of the sources of its unique electrochemical properties is hard to produce. A journalist commenting on scientific studies made the same point:

> There may be a scientific as well as religious basis for beliefs that this river can bring purification. According to environmental engineer D.S. Bhargava, the Ganga decomposes organic waste 15 to 25 times faster than other rivers. This finding has never been fully explained. The Ganga has an extraordinarily high rate of reaeration (the process by which it absorbs atmospheric oxygen), and it can retain dissolved oxygen much longer than water from other rivers. This could explain why bottled water from the Ganga reportedly does not putrefy even after many years of storage.[8]

Other syncretisms of Hinduism and science have been promoted by environmental activists (see chaps. 8 and 9). What these and other cases suggest is that theories of natural ecology have not attained a hegemonic status even among academics, journalists, and activists who might be expected to seek explanations for natural ecosystem processes through rational, godless categories.

TABLE 1. Aeration Rates of Several Large Rivers

	Thames	Rhine	Danube	Gaṅgā
Length (km)	245	1,320	2,857	2,535
Population (millions)		50	86	500
Restoration time (years)	30	50	13+	13+
Restoration cost (Rs billion)	5.0	1,940.0	125.0	11.2

Source: Ministry of Environment and Forests 1998. Status Report on River Action Plans, 21. New Delhi.

Religion and Politics—A Divergence?

Beyond the paper-and-pen strategies of scientists, academics, and bureaucrats, the intersections between religion and politics wear a different look and convey a less convergent and more divisive posture. Although religious rituals are used effectively to mobilize the voting public during state and national elections, political campaigns have not in an equally zealous manner attempted to exploit the message of Gaṅgā's purity in their convergences with religious discourses.[9] The first place to look for political appropriations of this message would be in the activities of the Vishwa Hindu Parisad (VHP), or the World Hindu Organization. The VHP stands out in the recent interchanges involving Ayodhya and in the politicization of religion in general. Almost fifteen years ago, the VHP picked up on an old conflict and expanded its metaphorical play by calling on Hindus in India and abroad to possess several sacred spaces located in northern India. The VHP argued that these were sacred spaces colonized by Muslims in the precolonial period. On the grounds of the Babri Masjid in Ayodhya, the Gyanvapi Mosque in Banaras, and the Shahi Idgah in Mathura, they vowed to avenge the past by destroying these places of Muslim worship, restoring them as sacred ground for Hindus and rebuilding Hindu temples. During the 1990s, the VHP joined hands with the Bharatiya Janata Party (BJP), a political party that had little strength in central and state governments at the time. With the support of VHP members, the BJP rather rapidly increased its representation in state and central governments and firmly assumed control of the office of prime minister in 1998 and again in 1999.[10] While combining their networks and the benefits of their spoils, the VHP and the BJP championed the cause of *Hiṁdutva* (Hindu nationalism).

During the early 1990s, as the VHP and BJP were stepping up their conflict in Ayodhya, political and cultural reporting on South Asia voiced concern for the potentially dangerous nexus developing between religion and politics. During this time, scholars revived debates on the nature of secularism and communalism.[11] But as the scholastic focus intensified, analysts of Indian politics, society, and religion began to highlight almost exclusively the processes bringing religious and political leaders together while ignoring the processes keeping these leaders apart. When looking into the differences between this religious-political movement to appropriate sacred space and win control of government and the activities of NGOs and government agencies to save or clean the Gaṅgā, several instances of the separation of domains become apparent. In what some might call a fledgling environmental movement, religious and political leaders have not been riding the waves of each other's victories or public performances in any concerted way. In fact, a more dissonant tone fills the air, even in the midst of possibilities for convergences. Curiously enough, issues involving the definition of river pollution in India reveal an unmined potency that even dissonance could use for its own purposes. River pollution carries a symbolic valence that both religious and political leaders could exploit for their own ends. An opposition political party could refer to the failures of the Ganga Action Plan as a way to attack the political legitimacy of the ruling party. Or religious leaders could mobilize their followers by pointing an accusing finger at Muslim tanners who dump industrial waste into the river. Yet these and other appropriations have not been made with any degree of effectiveness.

In 1995, I met an environmental activist who was asking VHP leaders in New Delhi to adopt a call to save the sacred river Gaṅgā from material pollution. This Kanpur activist was successful at getting VHP leaders to include this message in the agenda of a pilgrimage rally they held in 1995. However, during the actual processions, the message was never publicized. In effect a nonstarter, the issue failed to attract the attention of leaders of the VHP or members of other religious organizations. Apparently, leaders assumed that no political valence was latent in or could be created by the call to "save" or "protect" Gaṅgā from pollution. In chapter 10, I discuss why this was so.

In Banaras as well, religious and political leaders do not come together around the issue of Gaṅgā pollution. Residents remain largely disinterested in the government's first environmental program to "clean" the river. Even their anger against the government for spending millions of rupees on these projects is not a viable motive for mobilization. For Brāhmaṇ pilgrim priests (*purohit* or *paṇḍā*) who live and work on Daśāśvamedha in Banaras, the river Gaṅgā is a goddess who possesses the power to absorb and absolve human and worldly impurities. In a fashion cited in the *Viṣṇu Purāṇa*, they say she is able to wash over the impurities of the world like she washed over the foot of

Lord Viṣṇu on her descent to earth (*BhP* 5.17.1). Using her transcendent power in the contemporary context, Gaṅgā staves off the degeneration of contemporary society without defiling herself. They argue that since she remains pure and fertile, she should be worshiped by devotees and pilgrims through ablutions, meditation, and worship.

On the other hand, some Banaras residents refer to Gaṅgā's ability to purify as if to suggest that there is some kind of upper limit to her power. They suggest this when referring to her capacity *(ksamtā)* to purify or take away worldly dirtiness and impurity. Some residents acknowledge that waste is having an effect on the Gaṅgā, even when they cannot or will not define the nature of that effect. One way that religious leaders explain the possibility of an effect is by enforcing a separation between the domains of Hinduism and politics. In chapter 10, I explain that this does not reproduce essentialist oppositions between tradition and modernity or between this world and the next. This is because power, modernity, and this-worldly action already have a place in narratives of the religious and the transcendent. They are already encompassed.

PART 2

CHAPTER 4

Sacred Purity and Uncleanness

THE DEATH RITUALS PERFORMED in late 2001 by the relatives of the former Beatle George Harrison brought the unaware Westerner into contact with the purificatory powers of the river goddess Gaṅgā, and made the spiritualism of the East appear, again, as a kind of anomaly in the world of modern sensibilities. This complex confusion over what the Gaṅgā means to Hindus and what her relation is to purity, cosmos, and death weaves across continents. This confusion makes it all the more important to have a method for disentangling the various approaches to the Gaṅgā, for delineating distinctive discourses, and for excavating their intersections over time.

It is to the Hindu residents of the Gangetic Plain that I now turn, to discuss in greater detail how they express their understanding of Gaṅgā's power, dirtiness, and waste in the contexts of their religious rituals and everyday life. The discussion focuses specifically on Hindu residents and rather unfortunately excludes the perspectives of Muslims, Sikhs, and members of other religious and ethnic communities except as they participate within contexts I am calling scientific and secular. This is a limitation that will, no doubt, bring criticism, but it is one I was forced to make to draw boundaries around my focus and proceed in a systematic manner. The focus on Hindus takes the opportunity to look deeper into one particular discourse and find its intersections with other discourses. But I recognize that the very word *Hindu* essentializes many identities that individuals feel are more distinctive, such as those based on location, caste, or sectarian differences.

Parts of this text conform to a particular academic and cultural genre by virtue of the fact that they focus on problems of purity and impurity in the Hindu cultural universe (see chap. 2). This is a problem that has been central to South Asian studies for a long time (Appadurai 1988; Gupta and Ferguson 1997c). Following in the vein of this genre but at the outer wall of it, this chapter deconstructs the multivalent terms *purity*, *impurity*, *cleanness*, and *uncleanness* that are voiced by Hindu residents of a particular neighborhood in one north Indian city. These Hindus are residents of the neighborhood of Daśāśvamedha, which lies at the center of the city of Banaras (known as Varanasi in official discourse). These are residents with particularly close ties to the Gaṅgā in spiritual, economic, and political terms.

I use a method that cognitive anthropologists developed three decades ago to elicit semantic domains through structured interviews. Semantic domains are conceptual categories that act as frames for the signification of referents in the empirical world. They are domains of meaning that are invoked by linguistic terms or acoustic images, by what Saussure called *signifiers*.[1] In this chapter, I look at two particular semantic domains and outline what informants include in and exclude from these domains when they link them to referents in the empirical world.[2] Then I trace how informants create an overlapping semantic space to explain what happens when the two referents in their empirical world merge.

While taking this approach, I do not assume that a single taxonomy or paradigm can be built at the conceptual level to represent a cultural unity of some sort. Rather, I aim to show that informants from many backgrounds may use the same set of acoustic images or linguistic terms to signify a number of different semantic domains. In this chapter I concentrate on only one cognitive model and argue that it represents the perspectives of a network of Indian citizens who reside in one particular place. The following chapters show that through everyday discourse and the specialized discourses associated with the legal process, scientific inquiry, and official policy-making, this model intersects with other cognitive models to extend the number of semantic domains to which a limited set of acoustic images are associated. These semantic domains are invoked in debates about the Gaṅgā and wastewater.

It is to my informants that I defer on this point. They pushed me to realize that what is collective and shared about the meanings attached to and derived from linguistic terms varies according to many divisions in society, especially those of profession, caste, sect, class and gender. Indian citizens from different occupational, spiritual, and ideological positions talk about purity, impurity, cleanness, uncleanness, and pollution in divergent ways, but use some of the same linguistic terms to communicate their individual perspectives across caste, class, and gender differences. Therefore what Brāhmaṇ priests mean when they talk about purity and pollution differs markedly from what a government official means even when they use the same terms or the same terms transliterated into the English or Hindi languages. Since these linguistic terms can be associated with a number of different conceptual configurations, I would rather treat them as symbols than as the flat, framed end products of cultural exchange or cultural representation. I consider them vehicles, in the sense developed by Geertz (1973) and Herzfeld (1993), for the conveyance of many meanings and messages. The key purpose of using an ethnosemantic method is to separate the various concepts or meanings that these terms are associated with and relate them to a variety of discourses that speakers engage in. This approach should begin to reveal the crosscutting cultural and social positions from which linguistic terms are spoken.

But problems with ethnosemantic analysis remain. The tendency is to associate this method with an analysis that is removed from everyday politics and with the ways in which the basic definitions and labels for things are contested (see also chap. 1). By calling upon this method, I do not intend to disregard political intersections or to consider linguistic terms and the conceptual domains they signify as the only viable ethnographic data.[3] I would like to lead this examination toward a broader aim. By examining many kinds of ethnographic data—written documents, oral and textual discourses, ritual and institutional practices, and assessments of ecological conditions—I hope to get at the connections between conceptual, linguistic, discursive, political and ecological levels of order and process. Deconstructing the polyvalent terms *purity*, *impurity*, *cleanness*, and *uncleanness* is a way to begin piecing together conceptual fields that relate metaphorically and metonymically to larger social problems. The linguistic is one level in the process of symbolization (Ohnuki-Tierney 1981), a level at which sensory experience is translated into thought concepts that in a dialectical way signify, name, and label the empirical world. Entitling domains of meaning (Fernandez 1974, 1977; Parkin 1985) extends the application of key terms through metaphoric extension or what Fernandez (1974) called *predication.*

Focusing specifically on Hindus who reside in the Gangetic Plain necessitates a recognition of caste distinctions in everyday social practices but does not limit the discussion to terms that are related to caste-specific perspectives or that seem to function directly for caste interests. When interviewing and talking with religious leaders, pilgrim priests, ritual specialists, boatmen, and merchants, I assumed that their close connections with the Gaṅgā would meet my original interest in getting at the communities of resource users who know about each other's positions, practices, and voices. It is not necessary that the speakers recognize any solidarity that might emerge from their connections with Gaṅgā, although many of them do. Citizens involved in this sharing constitute a cross-section of the Hindu population and reside not only in Banaras but in other geographic locales. They are united not by sociological subgroup or geographic place but by their direct and often life-sustaining connections with the river. This sharing is the basis for relating their statements together as parts of a larger set of intersecting discourses.

At the beginning of my interview regimen, I tried to avoid prejudging the respondents on the basis of caste and gender. I did this because I could not predict at the time how caste identity and gender might prefigure their responses. At the initial stage, I followed several objectives while collecting what I thought would be good discursive data. My first objective was to find within Hindu discourses a semantic domain that had some commensurability with the notion of waste, a notion that Americans use to conflate trash, litter, and sewage (Kempton et al. 1995). This was not to initiate a comparative analysis but to separate out the conceptual underpinnings of my own assumptions. Through preliminary interviews, I found that many respondents used a common term, *gandagī,* to refer to the material conditions of waste. After conducting many interviews, I decided to choose this as a starting point for exploration but I did so with the understanding that it was not entirely similar to the American or to my own notions of litter, trash, and waste. The term *gandagī* coded material referents with a tentativeness that belied other important considerations, considerations that an American domain of waste would not generate. For instance, the American term *waste* lacks any relation to a sense of cultural place; it does not point to connections between place and power in the ways *gandagī* can. These are the metaphoric and metonymic connections that give the Hindi term its semantic breadth. To put it succinctly, it invokes ideas about geographic-sacred power, ideas that may also shape how it generates conditions of uncleanness and impurity. There are limits to the extent that it can impact or cause transformations in the human, natural, or sacred order. This is evident in informants' understandings of the relative power of gandagī. The Hindi term alludes to a vocabulary of place that relates to different kinds of power.

For Hindus living in the neighborhood of Daśāśvamedha, gandagī can be harmful to health and well-being, yet it is possible for sacred power to neutralize it. Thus the physical risks of gandagī are less serious when they are viewed in relation to transcendent power. While Americans tend to see waste as a problem that can be eradicated by modern technology (Kempton et al. 1995:65, 123–25), Hindus in Banaras subordinate the destructive power of gandagī to a sacred power that creates its own ecology. This is the sacred power that scientific theory attempts to naturalize. In their view, this sacredness transcends people and the powers of the natural ecologies they construct.

After deciding to focus on this term's semantic domain, I wanted to take my interview questions in a specific direction: to find out what residents of Daśāśvamedha think about gandagī in the context of its relationship to Gaṅgā. That is, I wanted to know how they would talk about the impact gandagī might have on Gaṅgā when it meets up with her. This, however, had to be put in a rather roundabout way, because many devout Hindus and religious leaders reject the notions associated with the term *river pollution* straight away. If I had prefaced the discussion by asking about how "gandagī pollutes Gaṅgā," discussions would have led nowhere. Instead, after agreeing with informants on what Gaṅgā's sacred purity means, I would begin a discussion of problems people have with gandagī. I began to see that real conceptual problems arose when interpreting the intersection of two substances—gandagī and the sacred river Gaṅgā.

At first, I had to find ways to ask about what they thought about a confluence of the two substances without referring directly to the conditions of uncleanness, impurity, or pollution. Since I could not suggest that "Gaṅgā was polluted," I asked, "When gandagī enters

Gaṅgā, what happens to her, to you, or to your religious practices?" Briefly put, many said, "nothing," then qualified that gandagī could harm humans and make Gaṅgā unclean in the short term, but in the long term could do nothing transformative to her. Wastewater, like other substances included in the semantic domain of *gandagī*, does not mix with Gaṅgā after entering the river stream but remains a separate entity, ultimately carried away by her flow. In this condition, if gandagī is harmful, it is so in only a very temporal way. By carrying away gandagī, Gaṅgā ultimately dissolves and overpowers it.

The question to ask, I soon realized, should have been put in the reverse: "What happens to gandagī when it runs through sacred space and enters a sacred stream?" This evoked all sorts of descriptions of the ways Gaṅgā is able to bring about material and spiritual transformations to the natural and human world that gandagī can not exert on Gaṅgā. To understand this further, I asked how the substances of gandagī might change after entering the Gaṅgā. At these moments, I seemed to be getting closer to an understanding of how informants were relating the semantic domains of gandagī and of Gaṅgā to the conditions of purity, impurity, cleanness, and uncleanness.

Most informants pointed to the following as elements of gandagī that meet up with the Gaṅgā: dirty water from drains *(nālās)*, industrial wastewater, organic and material trash, soap from bathing and washing clothes, human excrement from "doing latrine" on the riverbank, and spit from betelnut chew *(pān)*. Yet after entering the Gaṅgā, they are carried away or dissolved by her purifying power. In the process of being carried away or dissolved, Gaṅgā may enter a temporary state of uncleanness, a condition generated by this merging of substances. To describe these instances, residents of Daśāśvamedha pointed to a conceptual overlap between the semantic domains of *gandagī* and Gaṅgā (see fig. 4).

Apart from these remarks, residents also included "dead bodies" in the semantic domain of *gandagī* but suggested that the conditions produced by this merging were of a different order. Their remarks indicated that uncremated or partially cremated corpses are transformed by Gaṅgā's power in a different way. Boatmen and Brāhmaṇ pilgrim priests went beyond the materialist contours of gandagī by invoking and then fusing corporeal notions of flesh with sacred notions

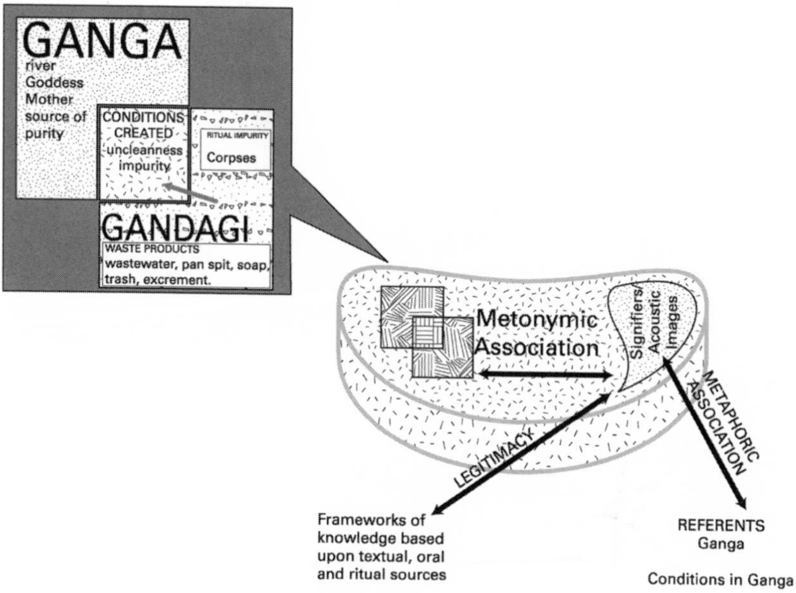

Fig. 4. Hindu semantic domains

of *mokṣa* (liberation) and *samsāra* (cycles of reincarnation). Uncremated or partially cremated corpses may create conditions of uncleanness or impurity, but this uncleanness is carried away by Gaṅgā's flow, and this impurity is converted by Gaṅgā's purificatory power. I will return to this distinction later in the discussion.

Further eluding a materialist approach, informants excluded from this semantic domain some things that a foreigner might call waste. Many residents of Daśāśvamedha did not include flower, food, liquid, or material offerings made to Gaṅgā. Not gandagī, these were considered blessed objects that must be immersed in the sacred vessel to complete ritual cycles of worship. The immersion of the flowers, food, liquids, statues of gods and goddesses, religious books, and saris is done with hopes for blessings, guidance, and benefits in this world, especially but not always at moments of cosmic juncture. To receive Gaṅgā's grace and power, pilgrims also bathe in the river and cook with and wash utensils there (see fig. 5). These offerings and practices regenerate cosmic order and please Gaṅgā immensely.

Fig. 5. Pilgrims washing utensils in the Gaṅgā

Gandagī at Daśāśvamedha

To examine the semantic domain of *gandagī* more closely, it is necessary to explore the views of pilgrim priests, boatmen, and merchants through their exact words, words that I and others have reproduced on tape cassette and recorded on paper. In these examples, I confine my discussion to interpretations drawn from male residents who work and live within the sacred space of Daśāśvamedha, in the city of Banaras.

Since I aim to argue that the term *gandagī* points to spatial and transcendent meanings more than the American terms *trash*, *litter*, and *waste* do, some background on space and divine power must be provided to set the context. Pilgrim priests and boatmen residing and working on Daśāśvamedha ghāṭ understand their space in terms of religious significances and everyday uses. Daśāśvamedha is an ancient, sacred place *(tīrth)* and a neighborhood of the city district of Varanasi (Banaras). It is also a sacred ghāṭ, a flight of steps that gives the public access to the river Gaṅgā (see fig. 6). Like the ghāṭs at Assī, Varaṇā, Maṇikarṇikā, Kedār, and Pañchgaṅgā, the ghāṭ at Daśāśvamedha is a site where the other world intersects with this world, where divine, eternal power can be felt and seen as immanent, and where the miraculous occurs in contexts of stark contrast. It is a place where divine power is accessed through elaborate ritual practices, guided and taught by local specialists. Finally, it is a place where Hindus acknowledge and complain about worldly interests in the midst of this otherworldliness and collectively deny their own interests in its disenchanting nature.[4]

According to sacred historiography, Daśāśvamedha is the ancient site for Lord Brahmā's ten-horse sacrifice. Through this sacrifice *(daśa-aśva-medha)*, Lord Brahmā gained the power to rule over King Divodāsa. The intersection of the power Lord Brahmā generated and the power provided by Gaṅgā is manifested in the temples of Śītalā, Ram and Sita, Śulataṅkeśvara, Brahmeśvara, the Gaṅgā Devī, and Bandī Devī, all located on Daśāśvamedha ghāṭ and on the contiguous space of Prayāg ghāṭ (see Singh 1993a:82). They mark the historic sacredness of the location, and they influence in an important way interpretations of gandagī's intersections with Gaṅgā.

The topography of Daśāśvamedha is also important to understanding the context in which local meanings of *gandagī* are formed. Most residents have an acute understanding of the intersection between the

Fig. 6. Daśāśvamedha ghāṭ (Photograph by Daniel Meadows.)

built environment and Gaṅgā, since they witness the cyclic increase and decrease in her flow, and they struggle with the physical transformations entailed by those flow changes. Her powerful flow, the silt she carries, and the levels she reaches must be watched constantly during the monsoon and tamed whenever possible by human efforts and material structures. These concerns are part of the basic ingredients of life along the river bank.

In its present constitution, the ghāṭ is divided into southern and northern sections (the Gaṅgā flows northward at Banaras). Prayāg ghāṭ lies between these sections at the spot where a natural drain (nālā) used to fork before entering the Gaṅgā. Flowing from a tank inside the city, this nālā bifurcated into two streams just before its confluence with the Gaṅgā, creating three sections of land separated by the two forks of the nālā. In 1740, the southern piece of land was paved with stone (made *pakkā*) by Bajirao Pesava I. Ahilyabhai Holkar of Indore extended the southern section in 1775 (Singh 1993a:82). In 1904, Maharani Puthia of Digpatia state in North Bengal constructed another ghāṭ on the piece of land lying between the two branches of the city nālā and named it Prayāg ghāṭ to invoke the image of the sacred confluence of Gaṅgā and Yamunā at Prayāg (now Allahabad). By the turn of the twentieth century, the principal roads of the city and the freight lines along the Gaṅgā converged at this ghāṭ (Havell 1990 [1905]:106).

Today the bed of this rivulet—once called the Godāvarī—is buried under urban construction. Yet the topography is still visible to the naked eye, especially during monsoon rains. During torrential downpours, the old bed fills with a fast-flowing stream of storm water that makes its way by gravity to the Gaṅgā (Singh 1993a:39, 84). It seems that this old rivulet once separated the sacredness of the space and made two sections of it. The area lying downstream from Prayāg ghāṭ—the northern section of ancient Daśāśvamedha—has grown popular in more recent times and caters to the many needs of commerce. However, ancient texts suggest that the southern section was the authentic site of ancient pilgrimage. It houses many liṅga and several shrines that are mentioned in sacred texts. On the southern section, we find the Śītalā temple, which contains the ancient liṅga of Daśāśvamedheśvara Mahādev, or Śivā as Lord of Daśāśvamedha (see also Vidyarthi et al. 1979:29). Behind this temple, the shrine of Prayāgeśvara lies underneath the house of a pilgrim priest. References to this linga and shrine are found in the *Kāśī Khaṇḍa*, a text describing Banaras between the eighth and thirteenth centuries (see Havell 1990 [1905]:110; *KKh* 61.36–38; Singh 1993a:48, 82–84).

The Brāhmaṇ pilgrim priests of Daśāśvamedha divide themselves into separate professions and use them to form a loose hierarchy. This is a hierarchy that many contest even as they claim to support it. Those at the top try to control rights over ritual services to pilgrims (see also Lynch 1996). These are rights that may be inherited, stolen, bought, or sold. The more powerful priests hold rights over large groups of pilgrims, especially over groups that have historically constituted a significant percentage of the pilgrim population. These groups are defined by their homelands, which are in many cases identified by names and boundaries recognized during precolonial and colonial periods. The ritual specialists of lesser power provide services to the smaller groups of pilgrims that arrive of their own accord and have no specific connection to a more powerful priest. Less powerful ritual specialists are often dependent on their superiors for access to the land on which they work, because they must rent the spaces where they sit on the ghāṭ from them. Therefore, they are always under the control of the priests who claim to own those spaces.

Three families occupy the highest stratum by controlling the rights to serve rather large groups of pilgrims from the state of Uttar Pradesh and the adjoining states of Madhya Pradesh and Bihar. The head male members of the family refer to their occupation *(peśa)* as *rāj purohit* or *tīrth purohit*, but they refer to each other and are known to outsiders as *paṇḍā*. One paṇḍā explains that the term originally referred to the middlemen who picked up pilgrims from the bus and train stations and led them to certain pilgrim priests. But the term has now come to stand for those powerful priests who are both middlemen and ritual specialists and who claim rights over large groups of pilgrims and arrange or perform ritual services for them. The term *paṇḍā* has taken on a derogatory image because these middlemen–cum–ritual specialists often engage in unsavory methods of maintaining control over rights to pilgrim groups (see Parry 1980:92; van der Veer 1988). To outsiders, the term *paṇḍāgīrī* (paṇḍā's work) stands for that which is grabbed, sinister, and calculating.[5]

Priests who rent spaces from paṇḍās are known by others as *ghāṭiyas* (those who sit on the ghāṭ), but they would rather refer to them-

Fig. 7. Pilgrim priest serves pilgrims on Daśāśvamedha ghāṭ

selves by what they consider to be the more enlightened term, *tīrth purohits* (those learned in rituals of the pilgrimage place). Generally, all pilgrim priests, whether paṇḍās, tīrth purohits, or ghāṭiyas, perform the worship rituals of pūjā, śrāddha, and tarpaṇ for pilgrims and, if they have the facilities, also provide basic amenities for the stay of more respected and wealthy guests (see fig. 7).[6] Ghāṭiyas may work independently by serving pilgrims on a walk-by basis (to those who arrive on their own accord), or they may work for superior paṇḍās. But overall, paṇḍās control the resources of Daśāśvamedha—the right to stand as guardian over tīrth purohits or ghāṭiyas and over merchants who sell commodities on the ghāṭ—because they claim rights over the larger groups of pilgrims.

Paṇḍās generally facilitate the movement of pilgrims from place to place, taking them to specific tīrth purohits, ghāṭiyas, boatmen, and shopkeepers. Pilgrims give fees, donations, or contributions of foodstuffs, clothing, and other valuables (*daksinā* or *dān*) to paṇḍās in exchange for shelter and the use of water and bathroom facilities, and to tīrth purohits and ghāṭiyas who undertake pūjā for them. A paṇḍā's earnings are supplemented by commissions that subordinate priests hand over after performing worship rituals for "the paṇḍā's" pilgrims.[7] All ritual transactions are based upon the fundamental assumption that rights are inalienable because they are inherited through divine grace. But this assumption belies the fact that paṇḍās have a persistent fear of losing these rights in contests for power and prestige.[8]

The three dominant paṇḍās of southern Daśāśvamedha claim rights over specific groups of pilgrims on the basis of hereditary ownership. However, paṇḍās argue that their hereditary right to serve pilgrims must be defended against shifting claims to resources made by individuals with "man and money power." This also figures into their understanding of power and the relative power of gandagī. But before going further into their notion of man and money power, I would like to turn back to the deconstruction of the semantic domain of gandagī.

The ways that paṇḍās, tīrth purohits, ghāṭiyas, and boatmen of Daśāśvamedha distinguish between physical cleanness and ritual purity are important to this domain of gandagī and to their assessments of the impact of gandagī on the river (or rather Gaṅgā's impact on gandagī). I include boatmen here because many of their remarks are quite similar to those of Brāhmaṇ pilgrim priests. Boatmen row wooden boats for pilgrims and tourists who wish to move along the waterway or simply enjoy the aesthetic beauty of the landscape and of cultural practices along the riverbank. Though boatmen tend to come from lower-caste families, primarily from Vaiśya and Śūdra jātis, their meanings of gandagī are quite similar to those of Brāhmaṇ priests because they are engaged in the same pilgrim/tourist business. This commonality of focus generates a certain amount of shared understanding about the power of Gaṅgā, even though their positions in other debates vary widely.

These ritual specialists and boatmen use eight terms, operating as four sets of binary oppositions, to explain the conditions created by this merging of substances: *sāf* and *gandā*, *svaccha* and *asvaccha*, *śuddha* and *aśuddha*, and *pavitra* and *apavitra*. The first two pairs—*sāf/gandā* and *svaccha/asvaccha*—are adjectives that refer to conditions of material or physical cleanness and uncleanness. The other two—*śuddha/aśuddha* and *pavitra/apavitra*—may refer to physical cleanness and uncleanness but more often they refer to purity and impurity of cosmos, soul, and heart.

Although residents treat them as sets of binary oppositions, these antonyms work in an ascending order of inclusion. For example, a resident might say that water is *śuddha* to mean it is clean and good to drink as well as sacred and purifying. This means that *sāf/svaccha* and *pavitra/śuddha* can signify similar conditions. In many ritual contexts, cleanness and purity are closely linked. Likewise, the same condition may be ritually impure (aśuddha or apavitra) and unclean (gandā or asvaccha). Hindu residents may use these terms interchangeably to refer to the same condition. But when focusing on discussions of the river Gaṅgā, they make this ascending order more apparent. In references to gandagī and Gaṅgā, informants indicate that substances that are pavitra or śuddha can also be sāf or svaccha but things that are sāf or svaccha are not necessarily pavitra or śuddha. The former concepts may encompass the latter ones but a reverse encompassment is not possible.

Paṇḍās do not say that gandagī is dangerous for Gaṅgā but they do value the rule for keeping dirtiness away from the sacred stream and other places of worship. They seem to draw inspiration from sacred texts and popular manuals on pilgrimage, spiritual life, and good conduct that communicate ideas about distancing unclean bodily functions from pure water bodies. A contemporary translation of the *Śivā Purāṇa*, for instance, makes numerous references to proper conduct near water bodies, and particularly next to rivers and tanks. About morning defecation and other routine activities, it teaches as follows:

> [For defecation], he must never sit in front of water, fire, a brahmin or the idol of any god. He must screen the penis with the left hand and the mouth with the right. After evacuating the bowels, the feces should not be looked at. Water drawn out in a vessel should not be used for cleaning (i.e. no one should sit inside the tank or river-water for cleaning purposes). No one shall enter the holy tanks and rivers dedicated to deities, manes, etc. and frequented by the sages. The rectum must be cleaned with mud seven, five or three times . . .
>
> . . . For gargling, the water can be taken in any vessel or a wooden cup; but water shall be spit outside (not in the river or tank). Washing of the teeth with any leaf or twig must be without using the index finger and outside the water. After making obeisance to the gods of water, the twice-born shall perform the ablution with man-

> tras. Sick or weak persons shall take bath up to the neck or hips. Sprinkling water up to the knees he shall perform the Mantrasnana. He shall propitiate deities, etc. sensibly with the water from the holy tank or river. A washed dry cloth should be taken and worn in the form of pancakaccha (wearing of the lower garment in a special way). In all sacred rites the upper cloth should also be used while taking bath in the holy river or tank; the cloth worn shall not be rinsed or beaten. The sensible man shall take it to a separate tank or well or to the house itself and beat it on a rock or on a plank to the gratification of the manes, O brahmins. (*SP* 13.10–13, 15–18)

These passages direct people to distance the everyday human activities of defecation, brushing teeth, spitting, and washing clothes from the riverbank. This principle of distancing, also mentioned in a favorite text of one paṇḍā, *The Path to Great Peace (Paramśānti Kā Mārg)*, appears to serve as a well-understood spatial benchmark for keeping uncleanness away from the Gaṅgā.

In his analysis of Indian wrestling, Alter also explains how the wrestling or *akhāṛā* ground reflects this spatial ordering. Banaras residents tend to distance uncleanness and human dirtiness from the center of the akhāṛā where a deep well of pure water draws its strength from the soil of the akhāṛā (considered a tonic of sorts), the trees, and its proximity to the Gaṅgā (Alter 1992:31–33). Wrestlers use the swampy ponds encircling the clean area of the compound for cleaning after defecation. The swampy area is the unclean periphery, where the dirtiness accrued to the body through everyday life is washed away. In this system of hydraulic classification, wrestlers distinguish between swampy water used to clean one's anus, water to dampen the ground, water to wash oneself and one's clothes, and water to drink. In this case, the purity of the akhara ground coexists with its physical cleanness, and both are distanced from physical and bodily uncleanness. On Daśāśvamedha, this rule of distancing is recognized as the ideal, even when it is breached. The collapse of the spatial ideal of separating what is unclean from what is pure disturbs most residents. However, they claim they cannot do much to change the situation. Most argue that the public nature of the ghāṭ at this spot makes regulation virtually impossible. Banaras, some point out, does not have a strong centralized religious authority, like that in the city of Haridwar, to enforce rules in a uniform fashion. On Daśāśvamedha ghāṭ, while pilgrims perform ablutions, others wash clothes with soap, a paṇḍā spits, and urban sewage flows into the river under the ghāṭ floor. Along the riverbank, early morning defecation occurs in a few open places. The vacant field upstream from Assī ghāṭ and the floodplain on the eastern bank of the river are the more popular locations for ridding the body of impurities. But many residents are quick to lament that some people defecate directly on the bathing ghāṭs (although generally in a less trafficked corner), spit the juice of their betelnut chew *(pān)*, and wash their clothes alongside pious bathers. Gandagī surrounds those who seek purification. While residents and pilgrims frequenting Daśāśvamedha may refer to the logical code of the śāstras to explain their ritual performances along the Gaṅgā, they admit that some break the code when it comes to urination or defecation and to popular habits such as spitting pān juice. The sharing of the river stands out as a distinctive feature of these human-ecology interactions (see chap. 1).

Tīrth purohits seated along Daśāśvamedha ghāṭ claim that they discourage people from using soap or oil while bathing, from using the ghāṭs as a public latrine, and from throwing dead bodies and animals into Gaṅgā. These are behaviors they dislike and consider religiously and socially incorrect. One tīrth purohit on Daśāśvamedha ghat explained in a taped interview:

> [TĪRTH PUROHIT]: Twenty years ago when my father used to sit there, twenty years ago till now, 80 percent of the people lived within an environment of vigilance. Now 60 percent do not live in an environment of vigilance. People come and they believe that this is *Gaṅgājal* (sacred Gaṅgā water) but while believing, while knowing, they do not believe and they do not know. This is the *bhāv* (emotion/sentiment) that has formed in their *man* (heart/soul), that when going near *Gaṅgājal* all the sins of the world are destroyed. While coming and taking such pure nectar *(amrit)*, at least they should keep the mud separate! Where is the intelligence in this? But while coming, people call it water, some believe it is *jal*, some understand it is *pānī* (secular word for water) and they corrupt their understanding in many ways: from entertainment, spitting, using soap, spitting, *nālās* from above. All this spilling into Gaṅgā and all these things make Gaṅgā dirty *(gandā)*. So the mentality of our pilgrims coming

today is corrupted *(duṣit)*, and the mentality of those men living on the banks of Gaṅgā is also corrupted. From the heart, they believe Gaṅgā is pure *(pavitra)* but by not believing in this purity, they think that Gaṅgā is very dirty *(gandā)*. They say: Gaṅgā is very dirty *(bahut gandā)*, we will get sick from drinking the water, from bathing; we may get heart sickness. Then from the scientific viewpoint, from believing *Gaṅgājal* is polluted *(praduṣit)*, this is how in the *man*, changes are born. Outsiders, Indians, and educated people have given a special importance to science. They agree that the Gaṅgā is affected by pollution and that Gaṅgā is polluted *(praduṣit)*. Our belief is that Gaṅgā *jal* on the side of the river bank is a little dirty *(gandā)* but the water in the middle is 98 percent pure *(pavitra)*. This is our opinion, our idea that water from the middle is not polluted *(praduṣan nahi hai)* in any way. There is pollution *(praduṣan)* in the water near the bank and this is caused by the people who live here and the pilgrims who come here. If the mentality should change and the feeling of humanity should change, like it was twenty five years before, if that feeling should come back, then Gaṅgā can be freed from pollution *(Gaṅgā kā praduṣan se mukt kiya ja sakta)*. If like this, soap will not be used, clothes will not be washed, *gandagī* would not be thrown into the *nālās*, dead bodies will not be floated in Gaṅgā, then the purity *(pavitratā)* of Gaṅgā will continue to increase. It will go on increasing.

ALLEY: If you see someone coming and using soap, do you tell that person anything like, "Do not use soap"?

[TĪRTH PUROHIT]: Yes, I tell them. While sitting on the *ghāṭs*, to the amount of people who live here, who go to the *tīrth purohits* and to pilgrims, we forbid them. We say you should not use soap in Gaṅgā because Gaṅgā is very pure *(pavitra)*. You come worshiping Gaṅgā as a Mother Goddess so why do you do dirtiness *(gandagī karnā)* near the Goddess for just one day? So please do not do this. Then people say, "I have come by travelling, I will use soap, this is Gaṅgājī's water, there is already pollution *(praduṣan)* in Gaṅgā water and here it is dirty *(gandā)* so a little more dirt will not hurt." Then we become silent. We are this silent because we have to take our donations from them. If we tell them in some special way not to use soap, they will not sit here but will run away. This is our employ-

> ment, our business, our work and this influences it. Therefore to prohibit this is difficult. From the side of the administration, if some order is made then people listen, for one–two days they listen. Then after this they do not pay attention. For one–two days they obey the administration—they do not use soap, they do not defecate or throw waste—but after two days, the administration changes and the work remains undone. After two days then people leave and those few who stay, they say: everything is OK. People come and go. This is what happens.

The sacred texts contain broad prescriptions to which everyday behavior cannot in every case conform. To meet the categorical ideal of separating unclean from clean, other facilities are required. Just as the swampy ponds serve as the appropriate site for the unclean but essential activities of urination and defecation for the wrestler, public areas ought to contain legitimate spaces for pilgrims to carry out these acts. But on Daśāśvamedha neither government authorities nor local leaders have made the required amenities available to the visiting population. Residents generally blame the government for the disorderly nature of unclean practices in places of ritual observance and then themselves for succumbing to bad habit.

Dead Bodies as Gandagī

Many Daśāśvamedha residents include uncremated or partially cremated dead bodies (called *laś* or *lāvāris laś*) in the list of elements of *gandagī* they spell out. Dead bodies have generally been considered carriers of ritual impurity and have been coded by the terms *apavitra* and *aśuddha*. But recently they have also been associated with conditions of uncleanness. When mentioning uncremated bodies that are dumped in the Gaṅgā, some residents conflate notions of ritual impurity and physical uncleanness to describe the conditions produced by this merging. Adding to its polyvalency, government officials, lawyers, NGOs, and foreign development agents consider these bodies signs of environmental pollution. To understand the ways corpses are associated with various arguments, several strands in the interpretive logics articulated by residents and government officials need to be separated.

The Hindu practice of cremation along the riverbank at the two auspicious ghāṭs of Maṇikarṇikā and Harīścandra aims to reduce the corpse to the five basic elements of existence: fire, air, water, ether, and earth. According to the views of pilgrim priests, deceased humans should be cremated, and after this the ashes should be immersed in the Gaṅgā to purify the soul (see Parry 1994). This constitutes a good sacrificial death, reaffirms Gaṅgā's purpose on earth, and enacts a sublime form of cosmic regeneration. Informants relate stories that tell of Gaṅgā's descent from heaven, a descent that brought her into contact with the three gods Brahmā, Śivā, and Viṣṇu. Through this contact, she acquired the power to purify human souls and absolve moral and material impurities. The *Bhāgavata Purāṇa* expresses this theme of imperishability in the parable of Gaṅgā's descent from heaven. Gaṅgā, who flowed over the foot of Viṣṇu and into Brahmā's jug, "washed away the dirt, in the form of the sins of the whole of the world, by her touch and yet remained pure [unpolluted by sins]" (*BhP* 5.17.1). After years of penance, a mortal named Bhāgīratha requested that Brahmā pour her out of his jug and onto the locks of Lord Śivā. Containing her powerful flow, Śivā guided her descent to earth (see chap. 3).

Hindus use the words *aśuddha* or *apavitra* to describe the condition of impurity brought on by cremation and tie these notions to ritual and cosmic order. According to Hindu concepts of death, dead bodies, per se, are not problematic for Gaṅgā because she has always accepted holy men *(sādhus)*, children, lepers, and smallpox victims, who are by rule immersed in a sacred river rather than cremated (see Das 1982:123; Parry 1994:184–85). However, paṇḍās of Daśāśvamedha point out that some corpses floating down the Gaṅgā fall outside this acceptable category. These particular corpses indicate a lapse in the public respect for ritual order and are considered matter out of place. The danger the corpses represent is not an ecological one but a moral and social one: rituals are being discarded.

Ritual impurity, then, is not always produced by uncremated corpses. This is explained in another way in Parry's (1981:360–61, 1994:186) discussion of two different beliefs about when impurity begins for mourners and the deceased. Among Banaras Hindus, one version holds that the corpse is ritually impure at the moment when vital breath leaves the body. The other version suggests that the condition of ritual impurity sets in during cremation at the moment when the chief mourner cracks the skull of the corpse with a stave. Parry (1994:188–189) explains that the latter version, while stipulating the onset of impurity within the cremation process, is linked to two beliefs, that the corpse is an animate oblation to the fire and that cremation is a good sacrificial death. From this position, impurity begins at the moment of cremation and is immediately followed by the purification process. But it is also possible that impurity may never inhere in an uncremated corpse if an effigy of the body is cremated. This suggests that the ritual of cremation, not the actual disintegration of the body, is the crucial purifier. The ritual impurity of mourners sets in at the moment of the corpse's or effigy's cremation, requiring Brahman priests to complete the purification process started by cremation. If we follow the first position Parry summarized we might be able to conclude that residents locate impurity in the drifting corpse. But if the other belief is held, the uncremated corpse floating in the Gaṅgā may only signify a bad death or the pollution of mourners and not, in itself, constitute ritual impurity. The corpse is a highly symbolic body, but not always ritually impure.

When residents include dead bodies in their accounts of what gandagī is, they may argue that these corpses create either impurity or physical uncleanness.[9] This means that the condition of ritual impurity coded by the terms *aśuddha* or *apavitra* and the condition of material uncleanness coded by the terms *asvaccha* or *gandā* are not mutually exclusive states. Both sets of terms may be used to signify uncremated corpses in the Gaṅgā, but only one term is used to describe Gaṅgā's condition. These bodies may render Gaṅgā temporarily unclean (asvaccha or gandā) but they can never make her impure (aśuddha or apavitra). However, both conditions are neutralized by Gaṅgā's purificatory power: impurity is converted into purity and uncleanness is carried away. Uncremated or partially cremated corpses can not push Gaṅgā into a state of impurity. Finally, dead bodies are significantly different from the ashes of cremated corpses that are immersed in Gaṅgā after cremation rites are completed. Although they represent an impure state, ashes have nothing to do with gandagī. Rather, they are impurities that are purified by Gaṅgā's power (see fig. 8).

The fact that some residents include dead bodies in the semantic domain of gandagī also reflects the influence that media reports of

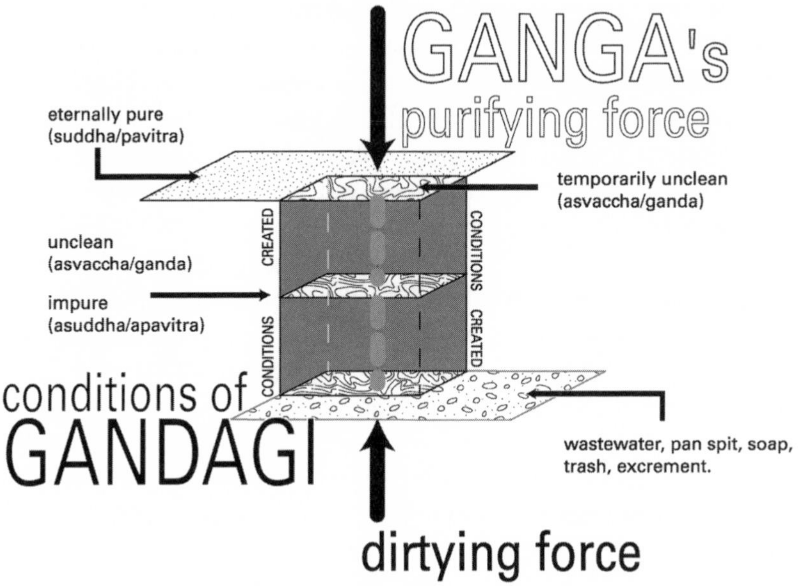

Fig. 8. Overlapping semantic domains of Gaṅgā and gandagī

environmental pollution have had on local thinking. Since the 1980s, media and official reports have marked these bodies as signs of river pollution and have warned about the visible increase in their number. Officials attribute the rise in the numbers of partially cremated and uncremated corpses in the Gaṅgā to the high costs of cremation, which make it difficult for some families to cremate their deceased kin. Media reports began publicizing this phenomenon in the late 1980s. From 1985 through 1990, reporters from Delhi and abroad published descriptions and photos of floating corpses in reports of "Gaṅgā pollution." Though they were meant to shock the citizenry into a concern for ecological degeneracy, they did little to convince Daśāśvamedha residents that Gaṅgā was in ecological peril.

Although paṇḍās know that media reports have used dead bodies as symbols of Gaṅgā pollution, they do not describe the conditions of uncleanness these bodies produce in the same way. This is evident when paṇḍās turn this value back onto those who do not believe in the sacred purity of Gaṅgā by pointing out dead bodies in phrases such as

"Look, Madam, dead body!" They expect the nonbeliever to be disgusted, outraged, and fearful for Gaṅgā. Such mocking exclamations made me realize that paṇḍās were less alarmed about an impact on Gaṅgā produced by these corpses than they were about social conditions. The decline in cremation rites disturbs residents, particularly those priests whose services are tied up with wood cremation. Priests are far more disturbed about the moral condition of humanity (and the implications this has for their own occupations) than about the conditions of uncleanness or impurity these corpses reflect. Some residents blame pollution prevention projects for the rise in cases of uncremated corpses in the river. Under the Ganga Action Plan, the government constructed an electric crematorium on Harīścandra ghāṭ to ease the pollution load on the river. This was to be a viable alternative to the costly procedure of wood cremation. Initially, pilgrim priests involved with cremation rituals opposed the project because it threatened to disturb Hindu practices of wood cremation (see Parry 1994:67–68). But the plant survived and now breeds consequences. Paṇḍās remark that dead bodies floating down the Gaṅgā today represent a decline in the respect for proper cremation practices, a decline set in motion when the government established an alternate form of electric cremation.

Paṇḍās firmly believe that these dead bodies do not threaten Gaṅgā's purificatory power. They insist that these bodies and all other forms of gandagī cannot alter Gaṅgā's power to give liberation (*mukti* or *mokṣa*) or to purify the ashes of the deceased. This power is eternal *(acyut)* and not subject to fluctuations in material reality. They add that as long as humans demonstrate their reverence through ritual ablution and other forms of worship, Gaṅgā will remain pleased. As long as she is pleased, she will continue to purify the cosmos, soul, body, and heart.

Looking more critically at this position, one could argue that a paṇḍā's voice on the matter also hinges in part upon his occupational interests, a point I will turn to in a moment. Yet I found that this voice also tends to hide more direct knowledge of wastewater flows that run under the ghāṭ floor. Paṇḍās are aware that these flows intersect with the Gaṅgā at the points where pilgrims bathe at their ghāṭ; because of this, two of the three dominant paṇḍās of Daśāśvamedha rarely bathe there. They recommend cleaner locations! The most powerful boatman in the area also complains that his doctor suggested he also avoid bathing at Daśāśvamedha, to ward off skin disease. He told me in a taped interview:

> I used to bathe in Gaṅgā. I also used to swim when my health was good. But now I bathe in my house every morning not in Gaṅgā nowadays because I am an allergy patient. So my doctor told me not to bathe in Gaṅgā, not to stay out in the sunlight. Cold Gaṅgā water can attack my body and give me an allergy. So to save my body, I have not bathed in Gaṅgā for three or four years.

Gendered Ecology

In *Water and Womanhood*, Feldhaus (1995) outlines how residents of Maharastra relate feminine imagery to rivers and stress the female attribute of fertility over the attribute of purificatory power. Banaras residents also think of the river in gender-specific ways, linking femininity not only with purificatory power but with motherliness, housekeeping and cleanup, and forgiveness. Many Daśāśvamedha residents and pilgrims have emphasized to me that Gaṅgā, like a good mother, cleans up the messes her children make and forgives them lovingly. In this way, she cleans up other kinds of dirtiness people bring to her and excuses dirty behavior with maternal kindness. Gaṅgā is forgiving rather than angry about human dirtiness.

As one elderly boatman put it, this motherliness can never be repaid because the same intensity of love cannot be achieved by a human individual. When asked if gandagī affects Gaṅgā, he went on to explain, "Gaṅgā is Mā. As Mā, nothing happens to her. Her water rises *(pānī bartī hai)* and the waste *(gandagī)* goes away."

Well, I continued, if there is no impact on Gaṅgā then why do people complain about gandagī? He replied:

> The ones complaining are saying that gandagī is coming from here and from there. Yes, they are acknowledging that gandagī is there, but where is the dirt in nectar *(amrit me gandā kahā hai)*?[10] When Gaṅgā is Mā then Mā will remain not the dirt. Rajiv Gandhi gave the call to clean *(safāi karnā)* Gaṅgā. But Congress [a political party] has

> not cleaned anything. In fact, they were the ones that were cleaned out [of power]! Who can clean Mā? She gave birth to humanity, so who will clean Mā? You can never repay Mā *(Mā se koi udhār nahi hotā)*. For example, there is one temple on Maṇikarṇikā ghāṭ. Mā Gaṅgā said to her son, "Make me a strong foundation and build me a temple." The son was a king so he built a beautiful temple. It is there today [on Maṇikarṇikā ghāṭ]. After it was built, the king went and did *pūjā* at the temple and then said, "From Mā we get life. She requested that I make the temple and so I have done it." But then the temple sank into the ground [and remains today tipped and sunken in the bed of the river]. So there is no one who can pay back Mā *(Mā se koi udhār honewālā nahī. Are, Mā hai!)*. In other words, there is no one to give her back all that she gives us.[11]

This view of Mother Gaṅgā speaks directly to her transcendent power to purify. But it also speaks about her immanent power as Mother. Her purificatory power is transcendent but her power as Mother pervades the everyday lives of humans. These are two sides of her essence that I elaborate in greater detail in chapter 10. She is so pervasive, so giving that humanity can never repay her, even by way of the most ambitious, scientific, and technological cleaning programs. When humans attempt to do so or to overpower or take over her role as provider and sustainer, they only end up in failure. Witness the sunken temple in the riverbed and the government's Ganga Action Plan.

This man's statement also demonstrates that he does not deny the presence of gandagī in the river. He and other residents of Banaras acknowledge this when they argue that Gaṅgā herself may become materially unclean (gandā or asvaccha)—that is, affected by gandagī—but not impure (aśuddha or apavitra). Like a mother who becomes unclean when her child defecates or makes other messes, she cleans others and herself without risking permanent defilement. Therefore when the semantic domains of gandagī and Gaṅgā are merged, several different kinds of conditions are produced, conditions that temporarily transform Gaṅgā and conditions that transform material substance. While material items such as uncremated corpses can be both unclean and impure, Gaṅgā can be temporarily unclean but never impure. Therefore like the robe Srinivas found (chap. 2), she is dirty but pure. However, if she is temporarily dirty (gandā or asvaccha), it is because people have made her that way (see Alley 1994:130). Humans, not Mother Gaṅgā, are to blame.

Gandagī and the Business of Daśāśvamedha

Environmental activists tend to think that residents of Daśāśvamedha passively accept the conditions created by gandagī by pointing to Gaṅgā's own power to solve the problem. But to understand the broader context for the positions just outlined, I need to trace how these connections between linguistic terms, semantic domains, and merged substances have something to do with occupational status and family identity. This can be done by looking more closely at how Brāhmaṇ paṇḍās connect Gaṅgā's purity (and purificatory power) to their own occupational interests. The paṇḍās of Daśāśvamedha consistently defend their conviction that human-created gandagī does not alter Gaṅgā's purity. This is a position with deep theological underpinnings. Yet one might also contend that this argument is a self-serving one because paṇḍās generate their livelihood from the purity of the Gaṅgā. They require a pure Gaṅgā in order to perform the rituals that pilgrims request, and they rely upon the fees and donations that pilgrims provide for their livelihood.

Pilgrim service can be a lucrative business. As one paṇḍā put it to me, "From sunrise to sunset, it is just earning, earning, earning." Many scholarly accounts have estimated that over the past four decades the number of pilgrims visiting Banaras daily to see the divine (take *darśan*) has steadily increased (see Fuller 1992:205; Parry 1994:108; van der Veer 1994:122). Local gossip puts the paṇḍās' earnings well above the average for Banaras residents. Two of the three paṇḍās I have introduced own substantial homes on southern Daśāśvamedha, and their residential dwellings are blessed by sacred shrines and icons. One mansion towers above the Rām temple on Prayāg ghāṭ. Another home contains the Prayāgeśvara shrine. The third paṇḍā reputedly owns a golden goddess. From lookout points on their property, these paṇḍās watch over activities on the ghāṭ and oversee the exchanges that lower-ranked pilgrim priests—tīrth purohits and ghāṭiyas—undertake with pilgrims.

Discussions with paṇḍās reveal the subtle ways that references to Gaṅgā's sacred purity are used for purposes other than worship to the divine. These pointers emerge from arguments about the limited impact gandagī exerts on Gaṅgā. Many residents of Daśāśvamedha argue that gandagī is problematic for Gaṅgā and not desirable for religious worship. But when asked if gandagī negatively influences religious ritual *(rīti-rivāz)*, many emphatically answer "no." While one would expect these statements more from pilgrim priests, I did notice that merchants, boatmen, and others involved in local commerce tend to speak with this voice as well. I attribute this to the fact that they are all engaged in the pilgrim service complex, a set of practices premised upon the power of ritual and communication with the divine. One ghāṭ merchant argued that because *sādhus*, *saṃnyāsīs*, *pandits*, *pujārīs*, and other spiritual masters keep coming to the banks of Gaṅgā for snān, gandagī has no impact on worship by reducing its validity, intensity, or viability. Admitting that pilgrims do not know, as their own family members do, about the hidden drains that run into the Gaṅgā at that sacred spot, they say pilgrims drink Gaṅgājal with complete faith. This faith is a great power that pandas and other service providers and merchants draw upon. But even if pilgrims know about the drains, many argue that they have no impact on her purificatory power. That power is eternal *(acyut)* and praised by the ritual practices of bathing, meditation, and worship. When asked whether Gaṅgā can purify all this dirty material without becoming impure, one panda replied: "Why not, she is Mā Gaṅgā! How much sin have we taken to her already!"

Alongside this faith in Gaṅgā's power, however, paṇḍās emphasize that human communities are caught in a period of moral degeneracy. To talk about this degeneracy, they also use the term *gandagī*, creating another conceptual frame for this familiar term. *Gandagī*, in these usages, refers to corrupt and immoral struggles for power and resources. But to flesh out this version of gandagī and connect it with the different orders of power that residents refer to, the basic structure of a paṇḍā's position of influence must be outlined. The locale remains Daśāśvamedha ghāṭ, not beyond it, since paṇḍās are very territorial and tend to stay within the limited sacred spaces they control.

As explained earlier, a paṇḍā on Daśāśvamedha maintains his position and gains power vis-à-vis other priests by controlling access to clients. The current paṇḍās have inherited their rights from their fathers, who, they argue, inherited rights from their fathers. Rights to act as head priest for specified groups of pilgrims may also pass through female lines, though these are less prestigious and are often disguised as gifts from the patriline. As local accounts go, many priests are unable to keep these rights within the family without struggle. As one informant explained by inserting the following English phrase into his Hindi dialogue, members of the community with more "man and money power" wrestle away titles from those who have inherited them. He described this form of "man and money power" as the reigning political mode on Daśāśvamedha. They connect this political mode with a more serious social malaise linked to cosmic decline. Pandas set the contemporary social and political conditions of gandagī within the Kali age *(Kali yuga)*, the last part of an aeon in the cyclical passage of the cosmos. Unlike other more poetic eulogies to Banaras that set the sacred city in an eternal *Satyā yuga*, an age of truth immune to degeneration (see Parry 1980:343, 1994:18), the pandas of Daśāśvamedha argue that Kali yuga is alive and well in Banaras.

In the sastras and in anthropological interpretations of Hinduism, the Kali yuga is characterized by diminished virtue, degeneracy, wickedness and sin (see Fuller 1992; Madan 1987; O'Flaherty 1976; *SP* 1990). Its beginning, like its end, is rather sketchy. Fuller (1992:266) sets the beginning at 18 February 3102 B.C. and its end 430,000 years or so later. However, informants do not provide reference to a bounded time frame (see also Madan 1987:128). They explain that the special characteristic of this period of ignorance and untruth is that many claim to practice severe and long-drawn-out austerities to win divine favor, but very few are able to reach the level where they can obtain these blessings. Voiced by paṇḍās of Maṇikarṇikā as well (Parry 1994:18–19, 49), this idiom invokes notions of a holistic decline ultimately beyond the full comprehension of believers and curious outsiders. In the context of Kali yuga, insistences about Gaṅgā's purity become ideologies that link cosmic theory to occupational positions and to the *khāndān* (family lineage).

In my discussions with them, the three paṇḍās of southern Daśāśvamedha rhetorically staged how they inherited and others stole away power in this field. Their knowledge has a temporal context, the Kali yuga, and a more specific frame of family memory extending back three generations to around the turn of the twentieth century. They trace their legitimate family rights to contracts made with *rājās* (kings of the former princely states). While two paṇḍās reckon descent through the male line, the third inherited his position through his great-grandfather, who took on rights through his affines.

As their accounts go, these rights were set for the three families of southern Daśāśvamedha by the kings of Palamau, Singrauli, and Sonbhadhra. After this, paṇḍā politics became more dynamic. In the early 1900s, Babua Pandey immigrated to Banaras from Bhalia and began to take over the paṇḍā business from two families, the Shuklas and Dubeys, that were controlling the area of Daśāśvamedha. Pandey built up "man and money power," as one informant put it, to gain control over the traffic in pilgrims throughout Banaras. Meanwhile on Daśāśvamedha the unity of the Shukla family was breaking down, rendering them vulnerable to Pandey's growing influence. Gradually, Dau Maharaj, a ghāṭiya who conducted pūjā on Daśāśvamedha, rose up from within the ranks created by Pandey and gained control over the traffic in Bengali pilgrims by the 1920s. He acquired control of the northern side of Daśāśvamedha ghāṭ and passed it on to his son and grandson who still work in that space. Remaining separated from these struggles, three ghāṭiyas working on southern Daśāśvamedha acquired rights from the three aforementioned kings and passed them down through the patriline.

After the 1920s, the paṇḍā business took a turn toward organized partnerships. Families controlling these and other ghāṭ spaces began to make alliances with middlemen who assumed the role of paṇḍā and guided pilgrims through the city. Over a period of forty years, Anjani Nandan Mishra, a prominent paṇḍā, rose to achieve a near monopoly of the business by making a nefarious alliance with an organization of middlemen known as the Joshi Brāhmaṇs. Brāhmaṇs of questionable origin (Parry 1994:107), Joshis developed a command over the pilgrims' arrival points and took control of their passage to and from sacred places. According to the few residents of southern Daśāśvamedha who retained their inherited rights amid these competitive shifts, the rise of Anjani Nandan Mishra and the Joshi Brāhmaṇs gave the paṇḍā position "a bad name." A general decline in authenticity began to develop, a decline that ultimately led to the contemporary state of social malaise. One paṇḍā explained:

> Joshi Brāhmaṇs were created by Anjani Nandan Mishra. They were born by him. From there they started. How? From the railway station, bus station, crossings. They used to stand there and from there they molded and cheated any pilgrims who happened upon that spot. They cheated them and took them to Anjani Nandan Mishra's ghāṭ. There they [ghāṭiyās of Anjani Nandan Mishra] did their [the pilgrims'] pūjās and Anjani Nandan Mishra gave a percentage [of his dāna-dakśinā] to the ghāṭiyās and to the Joshi Brāhmaṇs. Then after this, Joshi Brāhmaṇs became powerful and they thought: Anjani Nandan Mishra's business is going well because of me! Due to my society! Then they demanded that their commission be increased. But Mishra retorted: I created you! It is because of me that you are so developed today. Because of me! And you are always demanding from me? Then a fracture occurred. Fighting took place. Lathis were swung, murders committed. . . . After this the Joshi Brāhmaṇs moved away from there. Because they were so united and had so much money, no one could stop them. So in this occupation the powerful men are those who have man and money power!

Parry (1994) notes that the paṇḍās of Maṇikarṇikā ghāṭ, the ghāṭ on which cremation rites occur, also portray Anjani Nandan Mishra as the villain in their narratives about the degeneration of the paṇḍā position. Mishra stands for the extreme that dirty business reaches if paṇḍā politics is not tempered by the more preferable form of hereditary power. Yet even while paṇḍās maintain connections with Joshi Brāhmaṇs and other middlemen today, in public discussions they deny these relationships in order to cultivate a respect for inherited power. One paṇḍā mentioned more quietly that the Joshi Brāhmaṇs had been coming to southern Daśāśvamedha until a "fracture" occurred between the Joshis and the paṇḍā for Palamau. Now the Joshis take their business elsewhere.

The idea of "man and money power" is located here as an element of paṇḍā discourse. Paṇḍās invoke the temporal stage of Kali yuga to deflect the blame for dirty business away from themselves and toward an accelerated degeneracy lying beyond individual control. Calling themselves the "honest three" who struggle to survive amid the decline of their once respectable family service position, they hide the ways in which they defend their own rights to pilgrims. Indeed, as Parry

(1994:49) has noted, the language of material self-gain is as much a part of the discourse of paṇḍās as the language of purity and renunciation is, but self-gain is, in Bourdieu's (1977) sense, often misrecognized as honest, divinely ordained family service.

Paṇḍās lament that on other ghāṭs, paṇḍā work *(paṇḍāgīrī)* has become an occupation *(peśā)*, a profession *(vyavsāy)*, and a family business *(khāndānī vyāpār)* and is no longer a *jāt* (hereditary caste) position. People are now moving in and out of the occupation at will, and proving in the process that they can manipulate its nature. In its present state, the occupation lacks integrity and so do the activities of the people that paṇḍās are in charge of supervising. One paṇḍā continued:

> But all these things put together, *paṇḍā* has become such a word that we people have defamed and disgraced it; we have created a bad name for it. So we are ashamed! In society, you cannot say you are a paṇḍā because there is so much sin, senselessness, deceit, and cheating in this for the whole of India, for all the Hindu people at any tirth. I am not talking only about Kashi [Banaras]. You go to Prayag. There is also this occupation there. Go to Citrakoot. This occupation is there also. Go to Ayodhya. In every religious place or tīrth in this kind of way the business which goes on has already done enough ruin or ridicule.

Gandagī, in this sense, is a set of political circumstances that dirties the paṇḍā service role and casts a shadow over paṇḍās' power and moral authority to influence behavior toward the Gaṅgā on their ghāṭ. Dirty business means that as a paṇḍā competes to retain his hereditary rights, his moral authority declines. Consequently, pilgrims are cheated and Gaṅgā is abused. Yet while dirty business, cheating, and abuse reflect the sin of the Kali age, Gaṅgā continues to remain pure. As a paṇḍā's moral authority collapses into complaints about gandagī, the consumer interest in pilgrim services increases! The necessity for ritual specialists and for pilgrim services is maintained by the general view that purification, especially purification by an eternally pure Gaṅgā, is even more desirable in this age of degeneracy.

CHAPTER 5

Purity and Flow

> The reverence which the Hindu feels for the Gaṅgā and its water is equaled only by his veneration of the Brahman and the cow. He believes Gaṅgā water to be so holy that, let alone a bath or ablution, even a sip, touch or sight of it, washes away sin. He cannot however ascribe the same virtue to water drawn from the Gaṅgā into a canal which, in his eyes, does not possess the same virtue.
>
> *(Parmanand 1985:244)*

THE HINDU DISCOURSE ON GAṄGĀ'S POWER and the scientific discourse on natural ecology intersect over considerations of the river's unique flow characteristics. Hindus make this known when they point out Gaṅgā's capacity to absorb physical dirtiness and carry it away. As one paṇḍā of Daśāśvamedha put it, "Gaṅgā cleans herself during the monsoon" *(Gaṅgā khud hī sāph kartī hai).* Many described to me that, during the monsoon, Gaṅgā climbs up the ghāṭ steps and takes away the silt and gandagī. Following this line of reasoning, one merchant said, "From the scientific view, there is pollution, but I do not understand this. Gaṅgā takes the pollution away in floods."[1]

In the understanding that purity is connected with flow, Hindus have argued over many centuries that the river should not be contained, manipulated, or diverted by any form of technology. However, many residents of Daśāśvamedha today are not aware of the engineering accomplishments of the past century. The sentiment of current Banaras residents tends to reflect an older view of the river, a view that was expressed in the movement to oppose canal works at Haridwar in the early twentieth century. These sentiments are also held by some members of the Vishwa Hindu Parisad (see chap. 10). At the turn of the century, Hindu leaders insisted that Gaṅgā's purificatory power was tied to her flow and that to alter one would ultimately affect the other. They made these claims while opposing the colonial government's plan to divert a portion of the flow of the Gaṅgā from the main stream into an irrigation channel. This was to convert infertile into fertile land in the Doab and extend the reach of agricultural production.

Har-kī-paurī Ghāṭ

One sacred space stands out as a notable exception to the general rule that purity and uncleanness cannot be distanced in sacred places along the riverbank. This exceptional space is Har-kī-paurī ghāṭ, a sacred place located in the city of Haridwar. Haridwar lies just above the mouth of the Upper Ganga Canal and borders a branch of the Gaṅgā's sacred stream as it meanders through the foothills of the Himalayas. For Hindus, Haridwar is the gateway to earth for Mother Gaṅgā. It is here that she descended from heaven on the locks of Lord Śivā to relieve human suffering and purify souls. Kapilā and other early names for this place suggest that this site has been connected with Gaṅgā's descent from heaven for centuries. The Moghul chroniclers Abu Rihan and Rashid-ud-din and the seventh-century traveler Hiuen Tsang mentioned Gaṅgā-dvāra or the gateway to Gaṅgā.[2]

Historically, both Śaivites (followers of Śivā) and Vaiṣṇavites (followers of Viṣṇu) have worshiped at Haridwar and have developed different meanings of the city and its sacred spaces. Śaivites spell the city *Hardwar* and consider it the abode of Śivā (or Har), while Vaiṣṇavites call it *Haridwar*, the home of Viṣṇu (or Hari). Here I will use *Haridwar*, the official name of the city today.

Har-kī-paurī ghāṭ, situated on the western bank of the Gaṅgā, lies within the center of the sacred space of Haridwar.[3] At the base of the ghāṭ is Brahmā Kuṇḍ, a tank embedded in the river where Brahmā arranged the descent of Mother Gaṅgā to earth. Brahmā Kuṇḍ is filled with the nectar of immortality, the nectar that fell from the primordial jug the gods wrestled away from the demons.[4] The ghāṭ leading to this sacred tank is known today as Har-kī-paurī, Hari-kī-pairī, or Hari-kācaraṇ. Alternatively privileging Śivā and Viṣṇu, the first name means Śivā's door while the latter two mean Viṣṇu's foot.

Against the backdrop of this eternal sacredness, we find that the material shape of Har-kī-paurī ghāṭ has changed significantly over the past two centuries. In 1828, Hamilton described the ghāṭ as a narrow passage carved out of a steep mountain with "room for only four persons to pass abreast."[5] During the early part of the nineteenth century, several fatal incidents occurred in the narrow passage. In 1819, 430 persons including British sepoys were crushed to death as thousands pushed to bathe in the Gaṅgā on the most auspicious day of the *kuṃbha melā*, a festival that takes place once every twelve years at this site. After the stampede, the British government expanded the access way to 100 feet and increased the ghāṭ to sixty steps.[6] When the Sri Ganga Sabha was established in 1916, this organization took control of maintaining the ghāṭ for pilgrims and enforced rules for public behavior. Shoes, cameras, and cooking equipment were not allowed on the ghāṭ, and soap or oil could not be used while bathing in the stream. Later, the municipal committee constructed a cement platform, connected to Har-kī-paurī ghāṭ by two footbridges, to allow pilgrims to congregate on the other side of Brahmā kuṅḍ.

Hamilton's nineteenth-century description of the river landscape at Haridwar gives some detail on the nature of Gaṅgā's flow past this sacred site. At that time, the Gaṅgā flowed through several channels that lay a mile across an open gorge. One channel, departing from the main stream 2¼ miles above Haridwar, flowed past Haridwar and the pilgrimage places of Mayapur and Kankhal before rejoining the parent river. On this branch of the Ganga, the government of India harnessed the river's flow and created the Upper Ganga Canal. Modeled on the Eastern Jumna Canal built in the 1830s, Colonel Proby Cautley followed a path demarcated over two centuries earlier by Ali Mardan Khan, a minister of Shah Jahan, and charted out the route for the canal on horseback. The canal was to provide relief to a dry tableland that, lying at an elevation of 900 feet above sea level, was vulnerable to famine and drought.

After completing the survey work, Cautley was appointed by the government of the United Provinces to supervise the construction of the canal from headquarters in Roorkee. Cautley, an engineer and military officer, opened the canal for irrigation in 1855 and then ordered the construction of offshoots and distributors to widen its reach. But the colonial government did not begin to reap the rewards until over a decade later, well after Cautley had been charged with overspending. By the turn of the twentieth century, the Upper Ganga Canal had become the lifeline for the Doab.[7]

Negotiating the Canal Works

After the opening of the Upper Ganga Canal in 1855, engineering problems forced officials to close the plant every year for maintenance work. The canal head at Mayapur stood at a higher ground level than the main stream of the Gaṅgā did, and it produced a slower flow in the sacred stream (the supply channel) (see map 2). This allowed boulders and shingle to settle and cause shoaling at the headworks. After every rainy season in September and October, the irrigation department had to clear the boulders and shingle by blocking the entire flow of river water to the branch stream and to the irrigation canal. This caused problems for pilgrims visiting Haridwar and for farmers who required water for standing kharif crops (the crops reaped in autumn) and for sowing winter crops (Parmanand 1985:254).

In 1909, engineers of the Irrigation Department of Uttar Pradesh developed plans to create permanent works at Bhimgoda above Har-kī-paurī ghāṭ in order to regulate the flow of water in the supply channel. Without fully informing the public of their plans, officials assured the local paṇḍās that the works would ensure an increased flow of the Gaṅgā past the ghāṭ throughout the year. Only the All-India Hindu Sabha, an organization for Hindu nationalist unity, voiced resistance at that time (Parmanand 1985:245).

In 1914, two years after work had begun, the plan was radically modified to create a canal headworks at another site. The new plan included a proposal to build a masonry dam across the main stream of the Gaṅgā and construct a regulator with gates above that dam to divert water to the supply channel. To do this, a new channel had to be dug across Laljiwala Island. The irrigation department envisioned that this would prevent high floods in the original supply channel, reduce shoaling, and ensure an adequate supply of water for irrigation in the months of September and October.

Officials advocating this revision failed to anticipate how the Hindu

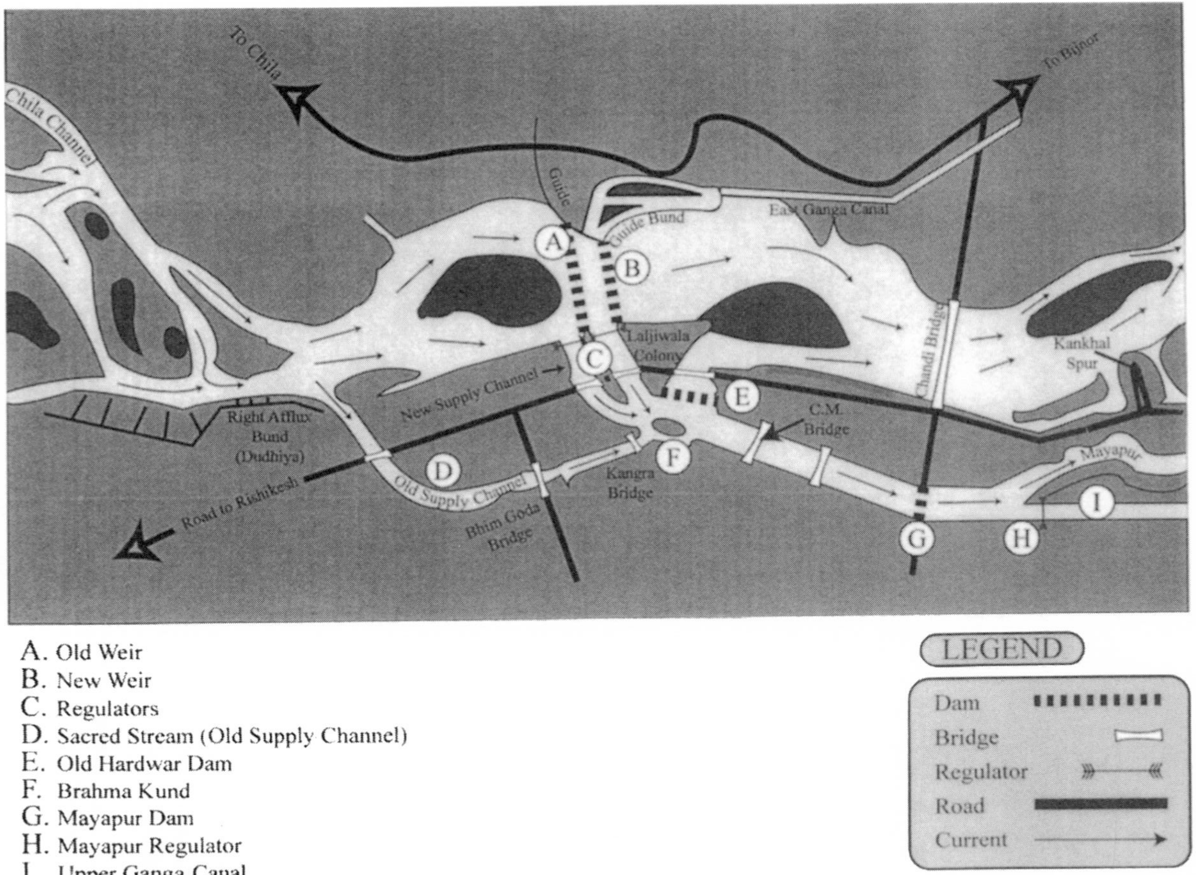

Map 2. The canal system in Hardiwar

community might react to obstructing the free flow of the river. Less appreciative of the benefits of canal irrigation, the public perceived a drastic shift in canal engineering and responded to that perception. In a public meeting at Haridwar on 5 June 1914, the local Hindu community expressed alarm at the projected works and demanded that the flow of the Gaṅgā past the ghāṭs remain uninterrupted (Parmanand 1985:247). Despite the public's response, the government of India went ahead and approved the expenditure for the revised plan.

Curiously, in its dispatch to the secretary of the state, the government of India did not spell out the details of the proposed canal works, but justified the increase in expenditure in terms of labor and machinery needs. Meanwhile, the agitation against these works, fueled by the demand for self-rule, gained momentum. Sensing this rise in public discontent, Sir James Meston, the lieutenant governor of Uttar Pradesh, convened a meeting of Hindu leaders on the 5th of November 1914 to discuss the issue. The mahārājā of Jaipur, a minister of the mahārājā of Alwar, the mahārājā of Darbhanga, Justice Chatterjee of the Calcutta High Court, Lala Sukbir Sinha, secretary of the All-India Hindu Sabha, Mahant Lakshman Das of the Sikh Gurudwara, Dehradun, and other public figures and engineers attended the meeting (Parmanand 1985:247–48). During this meeting, the representatives made a number of decisions and later released them in a communiqué. Under pressure from this Hindu opposition, the government agreed to keep the opening to the supply channel free from any blockade or sluice, even though some shoaling would result and have to be cleared annually. They also agreed to keep a "free opening" in the weir over the main stream of the Gaṅgā to allow for an "unobstructed supply" of water for bathers in the main stream. The Hindu community demanded that the new supply channel be cut "naturally" out of Laljiwala Island. Moreover, they demanded that engineers refrain from lining the canal with masonry, to avoid taking on "the name or the appearance of a canal" (Parmanand 1985:248).

The commitments spelled out in the communiqué were vague in several respects and avoided mentioning that work was already under way. Distrusting the government's assurances, Malaviya insisted that the mahārājā of Jaipur write a letter to Meston suggesting that plans to build both the regulator and the new feeder channel be abandoned. Subsequently, Malaviya and the mahārājā of Jaipur interested other ruling princes in the agitation. Another public meeting was held in Haridwar on 4 September 1916, at which time participants expressed their regret that the assurances given by the Irrigation Department in the 1914 meeting were not upheld. Copies of the resolutions from this meeting were sent to the mahārājās of Udaipur, Jaipur, Mysore, Travancore, Indore, Bikaner, Alwar, Gwalior, Jammu and Kashmir, Darbhanga, and the All-India Hindu Sabha. The Sri Ganga Hindu Sabha, a new organization representing the paṇḍās of Har-kī-paurī, also submitted their regrets to the government of India. In a separate correspondence, Malaviya and other mahārājās, rājās, former judges, and Śaṅkarācāryas brought their influence to bear upon Meston who communicated their concerns to the government of India. Warning the government against violating Hindu sentiment, Meston wrote:

> It is necessary to face the fact that Hindu sentiment was not consulted before these works were determined on, that it is extraordinarily powerful in all matters concerning the sacred stream at Hardwar and that the agitation—engineered and dishonest though much of it is—has penetrated and if unchecked will further penetrate vast depths of Hindu sentiment, which it would be most inexpedient to array against Government at any time and more especially at present. (Chief Secretary to Government of UP to Home Department of India, 5 February 1917, cited in Parmanand 1985:250)

The Hindu community pushed the government of India through Meston to abandon the construction of the regulator and other canal head works leading to the new supply channel above Har-kī-paurī. They also demanded that an opening of at least thirty feet be kept in the weir across the main stream of the Gaṅgā. Seen as irrational demands running counter to the needs for flood protection and irrigation, officials continued to disregard the protests. "Unobstructed flow" as Sir James Meston put it, was stretched to mean much that would be inconvenient (Parmanand 1985:252). However, Meston decided to accommodate the opposition by convening another conference in Haridwar on 18 and 19 December 1916. He invited the All-

India Hindu Sabha, the Ganga Hindu Sabha of Haridwar, and those who had attended earlier meetings on the issue. The mahārājā of Jaipur brought along his own contingent of ruling princes and their own technical adviser, a retired British chief engineer of the Punjab.

After taking a tour of the canal works, Meston sought the opinions of all those present. Their opinions were recorded and then used to formulate solutions. Graciously, Malaviya claimed that participants of the 1914 conference had misunderstood the government's decisions. He argued that they believed the new channel running from the regulator to the original supply channel would not bear the appearance of a canal. They also expected the weir across the main stream to have an opening without shutters to allow the free and unfettered flow of the Gaṅgā. The government had disregarded both these requests when they lined the canal and constructed a shutter in the weir. The Hindu community expected a larger opening in the weir, to permit enough water to maintain the river stream all the way to Allahabad. Malaviya pleaded:

> From the discussion which has appeared in the papers it is clear that people want a sufficient opening so that they can bathe. It must be wide enough to let a sufficient stream through. Thousands of people bathe along the course of the river. The opening must be so wide that all places below it will receive pure water. People seem to be dissatisfied with the Regulator. Possibly there was a misunderstanding both on the part of those who were present in 1914 and the outside public. It is important to remove this dissatisfaction in view of the holiness of the river. Even if some extra cost is incurred the feeling of the people should be soothed. It is said that the agriculturalists will suffer if the volume of water that passes into the canal is reduced, but no Hindu would place his material prosperity above the dictates of his conscience and his religion. In his opinion a five feet opening was not enough. Five to ten lakhs of bathers come to *tīrths*. They come from great distances and undergo great discomforts. But they will stand any trouble because it is a matter of faith. Even if the cost were one lakh or two lakhs, that should not matter when it was a question of belief with the people. This must be borne in mind. They believe that the Ganges makes people pure and removes sin. (Parmanand 1985:254–55)

Malaviya added that the works at Mayapur were supposed to allow river water to flow past Dakśā ghāṭ at Kankhal, but this, he pointed out, had not been achieved. The government strongly disagreed with the option of maintaining a permanent opening in the weir and regulator, citing the problem of shoaling and danger to the city from floods. Nevertheless, Meston was eager to forge a compromise and adopted a plan to keep a bay permanently open in the regulator. He also proposed to maintain a free opening in the weir to allow a minimum discharge of 400 cubic feet per second (cusecs) in the main stream during the dry seasons. In addition, the Mayapur regulator would have a free opening to provide a flow of 200 cusecs to the Kankhal ghāṭs. A minimum supply of 1,000 cusecs would be kept in the supply channel to feed the Haridwar ghāṭs.

While Malaviya was satisfied with the settlement, the Uttar Pradesh government and engineers of the Irrigation Branch felt that they had been forced to make concessions to a public ignorant of the facts of the situation (Parmanand 1985:257). Not surprisingly, in early 1917, the government retracted its accommodations while Meston was attending the Imperial Conference in London. They decided to divert the new channel from the regulator to a point on the original supply channel downstream from Har-kī-paurī ghāṭ. By redirecting the route of this new channel, they could argue that it would not mix with the sacred stream running past Har-kī-paurī. In March, the government of India authorized the UP government to inform the All-India Hindu Sabha that an alternative arrangement was being considered. However, withholding the announcement, the UP Irrigation Department proceeded to demarcate a line for the new channel on Laljiwala Island. When they noticed the work on Laljiwala Island, the Hindu community protested to Meston, who had by then returned from London. Meston wrote to the government of India to condemn the government's departure from the December 1916 agreement. The Sri Ganga Hindu Sabha of Haridwar followed with a similar plea. Surprisingly, on 20 September 1917, the government of India agreed to return to the 1916 resolution.

The 1916 resolution became a standing order for the Irrigation Department. Today, the executor engineer of the Northern Division Ganga Canal in Roorkee follows this order as it is reproduced in Appendix I of the *Standing Manual for River Ganga and Ganges Canal:*

I. To guarantee an uninterrupted flow of the Ganges through Har-ki-pauri and past the other ghats of Hardwar, Katchha (temporary) bund will be made at the head of channel no. 1 when necessary a minimum supply of 1000 cusecs in that channel being guaranteed except at periods when clearing of the shoaling in channel no. 1 is in progress; the Irrigation Branch undertaking that this work will be carried through as expeditiously as possible in order to ensure a flow from this channel into the Har-ki-pauri.
II. The opening to supply channel no. 1 will be left for the present exactly as it is. Should experience show that this is dangerous, it may be necessary to curtail the width of the present opening and to take measures to prevent the retrogression of the bed. But no steps beyond these will be taken without prior consultation with the Hindu Community.
III. The head of new supply channel will not be fitted with gates. Some bays contiguous to each other will be completely closed up with masonry and earth banks, some bays will be completely open. All bays whether open or closed will have a foot bridge. The floor on the Hardwar side will be level with the sell and with the bed of the supply channel. The Irrigation Branch reserves to itself the right of keeping open or closed by masonry as many of the bays as experience from time to time shows, may be necessary for the purpose of feeding the canal which starts at Mayapur and also in the interest of the safety of the town of Hardwar and the existing canal works but such bays as are kept open will be completely open as described above and whatever bays are closed, will be kept completely closed with masonry and banks as mentioned above.

The existing grooves will be left in what is now called the head of the supply channel. These grooves will never be used except in case of some impending calamity of accident to Hardwar or the existing canal works, when it is of vital importance to close completely the openings by means of wooden planks for a short period.
IV. A free gap will be left in the weir which will go down to floor level. The openings will be so constructed that it will give according to calculations of Irrigation Branch a minimum discharge of 400 cusecs at the cold weather low level of the river. For this purpose a record of gauges will be kept by the Irrigation Branch.

N. At Mayapur regulator a free opening going down to the upstream bed level will be made calculated to provide a permanent floor of 200 cusecs for the service of the Kankhal ghats, which after leaving Kankhal will ultimately flow into the Ganges.

N.B. The free gaps at Bhimgoda weir and Mayapur Dam may be closed by the Irrigation Department when necessary for the purpose of floor repairs or other emergencies but only for a short time and on informing the Hindu Sabha.

V. A depth of 4 ft. of water in the Har-ki-pauri pool will be guaranteed by and at the cost of Irrigation Department.

This agreement requires the government of India to provide a minimum of 1,000 cusecs in the flow of the stream that runs past Har-kī-paurī ghāṭ. This is to be an unobstructed flow, without barrage, gate, or shutter. However, the government is allowed to provide up to 11,000 cusecs, and the supply in the channel may reach 50,000 to 60,000 cusecs during the monsoon. During this heavy flow period, the two escapes—one across from Har-kī-paurī ghāṭ and the other adjacent to the canal headworks—are opened to help regulate the flow in the stream running past the sacred city.

Cleaning and Irrigation

At the sacred bathing ghāṭ in Haridwar today, the Sri Ganga Sabha enforces strict rules regulating public behavior. The Sri Ganga Sabha, which associates its origin with that crucial December meeting in 1916, is a trust set up to promote facilities for pilgrims and represent the interests of the paṇḍās of Har-kī-paurī. Originally established to protect the sacrality of Brahmā Kuṅḍ, the Sri Ganga Sabha maintains exclusive control of the management of the ghāṭ. The trust employs a staff of 150 to clean the ghāṭ daily and enforce rules for public use of the sacred space. Shoes and sandals cannot be worn on the ghāṭ, and photography, smoking, shaving, cooking, using oil or soap, and fishing are prohibited. The Sabha runs a free first-aid center and ambulance service for pilgrims, provides cold water in the summer months, and has volunteers on hand to assist the elderly and women with children. The trust manages one temple on the ghāṭ, and the others are run by paṇḍās on a hereditary or contract basis.

Across Har-kī-paurī lies a cement platform where pilgrims gather for ceremonies and recreational bathing. Although this platform is technically under the authority of the Sabha today, the municipality is responsible for cleaning it. On the platform, prohibitions against selling food are enforced, but shoes and smoking are allowed. The general secretary of the Sabha explains that regulation of shoes and smoking on the platform is too difficult because the platform has no sacred value. At Har-kī-paurī ghāṭ, on the other hand, pilgrims must leave their shoes with an attendant at the stalls bordering the sacred space.

The Sabha has pushed for and achieved a ban on liquor sales and meat eating in the entire city, as well as in the neighboring town of Kankhal. Its members form an effective pressure group when the local sanitation agencies of the Nagar Nigam are found lax in their work. The executive engineer of the Northern Division Ganga Canal Department pointed out in 1996 that the Sri Ganga Sabha is quick to raise "hew and cry" if the canal maintenance work is not carried out within the allotted time. Every year, from the day after Daśaharā to the day before Divālī (the festival of lights), the stream of the Gaṅgā that flows past Haridwar and Har-kī-paurī ghāṭ is diverted into the main stream so that repairs to it and to the ghāṭs at Haridwar can be made. Paṇḍās at Har-kī-paurī explain that the canal is closed from 5 October until the day before Divālī, because that is "traditionally the time for cleaning." The closing, however, is timed according to the needs of agriculture. This is the time that rice and sugar cane grown for fodder come to maturity. During the closure, the office of the Northern Division Ganga Canal in Roorkee repairs the canal, bridges, and regulators and removes boulders and debris from the supply channel. That office is responsible for the maintenance of the headworks of the canal and the stretch of the canal 83.6 kilometers beyond the source (Varun 1981:256).

The Sabha organizes an *āratī* (oil lamp) ceremony twice a day at sunrise and sunset. The evening arati attracts a large audience that gathers on the ghāṭ and along the platform across from the ghāṭ. Before the āratī begins, employees of the trust collect donations from the audience and provide a receipt from the trust. Donations made to the trust are the principal source of funding for all the Sabha's work and activities.

The current understanding of flow at Haridwar is expressed in terms of cusecs in the sacred channel. The Sri Ganga Sabha makes sure that the Irrigation Department maintains a level of 1,000 cusecs in the supply channel. This level has become crucial to the maintenance of the physical and material requirements of the sacred space at Har-kī-paurī. This level is required to make the physical infrastructure suitable to the organization of pilgrim traffic, to keep the space user-friendly and safe for pilgrims.

The conflict over the canal works had to do with defending Hindu interpretations of the meaning and use of the Gaṅgā against the scientific-colonial interest in exploiting the river for agricultural gain. The leaders of the resistance were not in favor of diverting the river at all and were especially against rerouting the flow away from Haridwar and Har-kī-paurī ghāṭ. In the early twentieth century, opposition to scientific control of the river aimed to defend sacred space and ensure the unobstructed flow of the Gaṅgā through it.

PART 3

CHAPTER 6

Defilement and Fouling in Colonial Law

THE SEMANTIC DOMAIN OF *GANDAGĪ* and the understandings of flow articulated by Hindu religious and nationalist leaders, pilgrim priests, ritual specialists, and service providers turn on their relationship to the field of sacred power. Moving by way of an outward layering methodology and turning to secular and official positions on waste, we find that the term *environmental pollution* is also quite complex in historical and discursive terms. This term is new to official-bureaucratic discourse in India, claiming a history of no more than thirty-five years in policy, law, and parliamentary debates. Colonial law and policy on water resources, on the other hand, offer an interesting set of linguistic antecedents, not because they provide for a neat evolution into the term *environmental pollution* but because they set up official positions on the relationship between water and caste politics. In colonial law, we find that the terms *contamination*, *defilement*, and *fouling* were used to settle cases on well water disputes. These cases provide a window into the ways in which, from the beginning of government policies on water resources, the concern for water potability ran antithetical to caste notions of impurity. This concern underlies the official-bureaucratic discourse on environmental pollution and river pollution in India today.

The starting point lies in the discourses of colonial officials, who used the term *physical defilement* to code specific human interactions with water resources. Colonial officials' approach to the relationship between sacred and secular power differed from that of Hindu religious leaders and ritual specialists (see chaps. 4 and 5). In fact, they attempted to enforce a strong distinction between the two and incorporate that distinction into the civil and criminal laws they were creating. They did this to assert the primacy of civil over divine power. Later, as postcolonial law took shape, official discourses moved the terms *defilement* and *contamination* in other directions.

Quite surprisingly, legal discourses on resource use and resource management have not been central to studies of political ecology.[1] Even poststructuralist analyses in political ecology have not accounted for legal discourses on resource use rights and battles. If we look at social science research on environmental justice in the United States, we find that this field has had much less to do with legal discourse and particular environmental cases than with cultural norms of justice and breaches of those ideals.[2] It may be that studies in environmental racism have not accounted for legal defenses and arguments because there have been few chances for grassroots groups to argue in their own defense or arrange for counsel to defend them in court (though such litigation and studies of it may well increase in the coming years). Limitations to citizen standing (see chap. 9) also make this a marginal part of the U.S. judicial process. In some other countries, however, public interest litigation is widely recognized as an avenue for problem solving (see Kaniaru et al. 1997).

An examination of the terms *defilement*, *fouling*, *contamination*, and *pollution* in legal discourse offers an important starting point for understanding the evolution of official environmentalism. In the colonial period, laws on resource use directly reflected British interests and cultural dispositions. Examining colonial laws on well water and other water resources shows how interests in expanding colonial hegemony were executed through specific semantic shifts meant to bring about a slow change in the government's approach to important water resources. Justices facilitated these semantic shifts in and through the political and economic appropriations of resources that other colonists were undertaking at the time. These semantic shifts guided and confused understandings about transitions from common property or caste-based regimes of water management to regimes of private and state property. They also show that legal discourses have been crucial to the construction of a theory of natural ecology that now gives water resources a distinctive secular character.

Many accounts of environmental history in India have focused on the use of forest resources in precolonial and colonial periods, especially in the Himalayan range.[3] Apart from this interest in forests, there are few if any accounts of the precolonial period that provide evidence for rules about water use or information on penalties levied against individuals who degraded water sources (Rosencranz et al. 1991:28–29; Thapar 1973:264). This rather curious lacuna may be due to a loss of material evidence over time. By contrast, colonial records reveal an emerging concern with water resources and with protecting them from contamination and degradation. By the end of the nineteenth century, the benefits of the country's water resources were more clearly defined; they assumed a place alongside concerns with forest resources. Reflected in the enlarging scope of law and opinions on well water disputes, this focus evolved into policies regulating open-access and common water resources. From wells, the interest expanded to sea harbors, irrigation canals, rivers, and oceans.

The data presented here demonstrate that historical transformations in the uses of natural resources are complicated at the discursive as well as the political, economic, and ecological levels. But while competing strategies for the utilization of scarce resources may suggest something about these complications, they do not tell the whole story. We still need to know how specific concepts and terms are appropriated to serve different discursive purposes over time (see also Brosius 1999b). When looking at the semantic history of the terms *defilement*, *fouling*, *contamination*, and *pollution* in India, the entanglement of official and native agendas is evident. This entanglement is visible through a series of internal contradictions that emerged in colonial laws and reports. It is not at all clear that the naming of conditions allowed officials to physically control scarce water resources or the properties on which they rest. And even though they had more to do with what Foucault (1991) called the technologies of objectification that built a governmentality, it is not at all clear that they achieved their hegemonic effect.

These entanglements began to surface as colonial authorities drew up a legal code to apply across the subcontinent. With this agenda in mind, colonial officials set out to manipulate village rules on local resource use as a way to slowly usurp other resource management tasks. The British used the domain of law to expand their power over natural resources grounded in the soil of particular communities and controlled by the power holders within them. But initially, colonial law took shape at the periphery of these local forms of control, operating in the interstices of local power. Data on disputes over well water reveal some of the ways in which colonial authorities tried to complicate indigenous rules on the management of water resources. Colonial laws sought hegemonic power by shifting the focus of village debates about access rights toward a concern for water potability. Unable to appropriate control over property rights to village wells, colonial officials proposed another way of controlling crucial water resources and devised a legal justification for it. This was a justification that eventually swept beyond wells to larger water bodies and flowing rivers.

Nonetheless, this strategy reflected the colonists' limited powers. Colonial officials could not coopt use rights to wells because these rights were firmly embedded in the village soil and tied to local politics and claims to land. Taking a different tack, they tried to control knowledge about well water and use this rationality to settle disputes about access rights. This knowledge came to exert its own disciplinary intentions and normalizing mechanisms, to bring well water under the administrative apparatus of the state.

Colonial laws on water corruption, fouling, and pollution confronted resistance at the village level. Embedded in the digests are indications of the ways native Indians were defining and practicing rights to resources and well water at the time. However, several caveats need to be made. Without primary data, there are at least two problems with reading the indigenous through colonial frames. First, any interpretations of indigenous power that might be gleaned from such an analysis run the risk of essentializing the Hindu village and the Hindu reality. Rocher (1993:221–22) explains that the colonial government lumped many religious communities under the label *Hindu* and treated them under the purview of Hindu law. There are very few sources of data, if any, that suggest the views of Jain, Sikh, Parsi, or Muslim participants in disputes over well water. The other problem with reading colonial law as a way to understand indigenous rules of resource use follows the charge that colonial law comprised a world of supralocal, symbolic domination that was divorced from Indian life and did little more than legitimate itself (Galanter 1989:3). As Galanter has argued, the normative intentions of law prevent the reader from understanding the everyday practices affecting those codes.

These problems can be addressed in the following ways. First, the normative nature of law need not frustrate the goal of reading beyond rules if there is no expectation that these rules had exact applications. As Galanter points out, they are socially constructed rules reflecting the rationalizations of the parties' own actions in disputes and only incorporate a "measured view" of the context at hand (1989:11). It is necessary to cull out of them clues to how other vectors were injected into the resolution of disputes. Wider political processes implicated in these legal documents played a role in muddying up the outcomes of settlements. This means that legal proceedings of dispute settlements must be treated as allegories of a deeply dispersive nature.

Legal cases from the colonial period provide the most accessible window on the government's policy on scarce water resources. Since there are very few cases to work with, references to well water disputes in law digests must be fully mined. This paucity of data does not mean that disputes rarely existed in the law courts. Rather, Galanter (1989:5) reminds us that cases recorded in law reports only represent the last stage of legal proceedings and usually the last stage of appellate cases. What has been documented stands for only a portion of legal history and a particular stage in the resolution of selected cases. So the surviving evidence does not represent accounts of trials in which witnesses give testimony and evidence is analyzed. Rather, the recorded opinion tends to tell how the subordinate courts acted under given circumstances and then records the decisions of the superior court. Law digests, while tending to show how dispute settlement is steered into decisions on specific issues, do not outline the shape of larger disputes. Moreover, opinions say very little about the implementation of judicial orders and laws at the village level. Given these problems in representation, context, and applicability, let us find out how decisions on well water disputes can be read for their effects on policy positions over time.

Defilement and Caste Impurity

Colonial law sought to undermine customary codes but did so in a rather roundabout way. Without directly challenging the rights of certain villagers to well water, they directed their concern to the definition of water quality and expressed this in discussions about the specified uses put to well water by individuals. Cases involving village-level disputes over the use of well water show that colonial authorities sought to undermine indigenous codes of resource use in contradictory ways. Quite often, justices were not consistent in the reading and application of the laws on their books, especially those laws that sought to undermine caste-based controls and powers. These contradictions were produced as justices connected and disconnected the notions of caste impurity and physical defilement in orders and opinions over time.

Contradictions in colonial law can also be seen in terms of the uneven way in which colonial officials defined and applied law. The first judicial plan prepared in the late nineteenth century proposed that Hindu law apply to Hindus in matters that dealt with religious usages or institutions (Galanter 1989:21–25; Rocher 1993:220). All other matters were to fall under the purview of common law. As Rocher (1993:220) points out, this put the British on the dangerous course of having to discriminate the religious from the lay under changing conditions. Officials decided quite early that cases involving water sources would fall under common law and be treated in "secular" terms. Early on, colonists attempted to take water disputes out of the domains of Hinduism and casteism.

The Indian Penal Code provides an interesting set of provisions to analyze for information on the colonial government's position vis-à-vis uses of well water. The Penal Code was one of many legal codes that British officials drew up in the mid-1800s. Based on English law, these codes covered the fields of commercial, criminal, and procedural law. Leaving separate Hindu and Muslim personal laws, the codes were entirely British and represented no fusion with indigenous rules and traditions. The Penal Code set the criteria for crimes and criminals and also the parameters for assessing other factors that shaped the definition of criminality.

The cases outlined here have less to do with criminality and more to do with an evolving definition of natural ecology. Though described in the cryptic style Galanter warned about, the cases mentioned in the law digests tell us something about the evolution of a theory of natural ecology, one envisioned in direct contradistinction to other theories about human interactions with water. As mentioned earlier, cases involving definitions of well water and its natural ecology suggest the forms of resistance colonial agents encountered. Colonial attempts to regulate well water imply that wells were controlled by another regime of power, a regime connected to caste politics.

In 1860, the Penal Code set out the government's position on village well water. It read:

> Whoever voluntarily corrupts or fouls the water of any public spring or reservoir, so as to render it less fit for the purpose of which it is ordinarily used, shall be punished with imprisonment of either description for a term which may extend to three months, or with fine which may extend to five hundred rupees, or with both. (Act XLV of 1860, cited in Mathur 1980:90; see also Manohar and Chitaley 1979c:268)

This statement aimed to punish those who voluntarily corrupted or fouled water used by the public. However, the statement was ambiguous on three counts. First, the clause embodied the term "voluntarily" to suggest that to be criminal, behavior must be motivated and intentional. Second, the paragraph proved problematic because it elided the distinction between caste impurity and physical defilement, a distinction that the British judiciary soon found itself entangled with. Finally, the statement made no mention of any distinction between wells, reservoirs, and rivers.[4]

The legislative digests briefly describe several cases in which petitioners and justices made distinctions between caste impurity/pollution and physical defilement in disputes over well water, to confuse the original spirit of the Penal Code. The ambiguity in the Penal Code made it possible for justices to interpret conflicts at the village level in different ways for different purposes. This is evident in the inconsistent ways in which they applied the terms *corruption*, *fouling*, and *degradation*. Section 277, for example, defined the "corruption" or "fouling" of the water of any public spring or reservoir as an act that renders the water less fit for drinking purposes. In a commentary on Section 277, the digest to the Penal Code mentioned:

> In the context in which the section occurs, the expression "corrupts" or "fouls" in the section must be taken as referring to the corruption

> or fouling of water in a physical sense so as to make it unhealthy or offensive to the senses and not merely in a sentimental sense so as to offend the religious sentiments of a particular community of persons who regard themselves as belonging to a higher caste.[5]

Several years later Section 295 was written to admit some flexibility into other readings of corruption and fouling. It was created to prohibit the defilement of any object held sacred by a class of people. While Section 295 was used in cases involving sacred objects or places, Section 277 became the position on defilement used in well water cases. Section 277 imposed an English view of defilement, while Section 295 accommodated Hindu notions of caste impurity and sacrality into its definition of fouling.

The differences between sections of the Penal Code become evident in the rulings made in several cases. In a case filed in 1900, justices invoked Section 277 to settle a dispute over well water. In that case, a lower-caste woman was convicted under the District Police Act of 1890 for corrupting water in a public cistern, making it less fit for drinking purposes. The District Magistrate referred the case to the Supreme Court so that the conviction under the Police Act could be withdrawn and replaced by a trial under Section 277 of the Indian Penal Code. The law digest summarized the conflict as follows:

> The Magistrate said, "The accused is by Vanjari caste [by caste a Vanjari]. Water can be taken by high caste without any objection being raised by Marathas or Kunbis. But she being kept by a Ramoshi, the water is not taken by the higher caste. The objection was raised by other people, but accused in contravention of repeated directions not to touch it she purposely did so." The Magistrate, therefore, convicted the accused as shown above, and fined her eight annas, which fine was paid.[6]

The Supreme Court rejected the plea to retry the case under the Penal Code and ruled as follows:

> We are unable to accede to the District Magistrate's suggestion. We think that the defiling or causing to be defiled the water and c. referred to in Section 61(m) of Bombay Act IV of 1890 and the

> voluntarily corrupting or fouling the water and c. referred to in Section 277 of the Indian Penal Code both mean some act which physically defiles or fouls the water.
>
> The District Magistrate says that accused, in direct contravention of the objections raised, intentionally fouled the water of a cistern intended for people of high caste and therefore she committed an offense under Section 277, Indian Penal Code. As far as we can gather from the papers submitted, the act of intentionally fouling was the act of drawing water from the cistern. It is not pretended that in fact the water was physically rendered less fit for the purpose for which it is ordinarily used. It is not shown that the "repeated directions" or "objections" of people had any legal force. If they had, accused might possibly have been prosecuted under another section of the Indian Penal Code. Under these circumstances, we return the record and proceedings.[7]

In this case, the Supreme Court rejected the notion that the woman's caste identity could play a legitimate role in her motives for using the well water. She did not intentionally foul water simply by attempting to use it as a lower-caste person. Further, the judge ruled that the objections by upper-caste villagers held no legal force. In effect, the Court argued that traditions based on accusations of caste impurity, traditions that restricted the use of some village wells to persons of higher caste, held no legal basis. An individual's caste impurities alone could not physically defile the water and render it undrinkable for others. In this case, the justices removed the caste identity of the user and the condition of caste impurity/pollution from the legal definition of corruption or fouling of well water.[8]

This ruling can be read as an attempt by British officials to undermine sanctions based on caste identities. But this was a tricky intention, one that undermined caste discriminations in some conflicts and supported them in others, to produce a rather contradictory orientalist vision (see Breckenridge and van der Veer 1993; Dirks 1992; Rocher 1993). This contradictory vision was in part shaped by the fundamental distinction officials had made almost a century earlier between the ecclesiastical and secular domains. In their view, the two domains were to be separated, and disputes over well water were to be tried under common and not Hindu law. Originally, the British interest was to hold Hindus to Hindu laws in contexts conducive to such interpretations, but this intention assumed a different shape over time. A century later, it had produced a more amorphous aggregate of policies that pushed and pulled caste-based customs in different directions.

In the midst of this, some consistency is evident in cases regarding determinations about the potability of water. Again these were determinations colonial agents associated with the emerging secular domain, and they were decisions over which they sought to exercise wider control. Landowners and upper-caste groups generally held use rights over wells and decided use rights for the public. Colonial authorities did not contest the use rights that were pegged by village power holders to caste identities. Instead, without directly challenging use rights or the discourse on use rights played out in disputes about well water, judicial officials deflected the focus of dispute settlement to the determination of potability. Indirectly they tried to delegitimate arguments about caste-based determinations as they began to define and sanction the state's role in deciding standards for drinking water.

This shift to the secular domain is also evident in a case heard in 1882. In that case, a low-caste group of Mahars was accused—presumably by members of an upper-caste group—of drawing water from a well. They were accused under Section 295 of the Penal Code. Referring to Section 295, the Court ruled: "The drawing of water from a road-side well by Mahars is not an offence under this section as the well is not an object held sacred by any class of persons" (Manohar and Chitaley 1979c:304). The justices relegated this issue to the secular domain by refusing to apply Section 295 (that is, the more accommodating Hindu interpretation of defilement). They were consistent here in removing considerations of caste status and impurity from evaluations of water potability and use rights.

In cases on defilement that did not involve well water, officials were less consistent. In several other recorded disputes, they used Section 295 to meddle in the ecclesiastical domain in a more direct way and settle disputes on the basis of caste distinctions. The wording of Section 295 made it vulnerable to meddling of this sort. It read:

> Whoever destroys, damages, or defiles any place of worship, or any object held sacred by any class of persons with the intention of thereby insulting the religion of any class of persons or with the

> knowledge that any class of persons is likely to consider such destruction, damage, or defilement as an insult to their religion, shall be punished.[9]

A reference to this section in the All India Reporter stated: "'Defile' is not confined to the idea of making dirty but is also extended to ceremonial pollution, which however must be proved."[10] This section in effect established that the Court was willing to make a distinction between physical cleanness or uncleanness and caste purity and impurity under what they saw as "ecclesiastical" conditions. Moreover, they were willing to defer to religious powers in the settlement of disputes over these kinds of defilement. In a commentary on Section 295, one justice added that defilement depended upon custom and the rules of a particular temple. If by custom, a temple had forbidden entry to untouchables, then the section would apply. If a temple had customarily permitted all castes of people to enter, then defilement by caste status could not be argued in a court of law. The law put decisions regarding access to temples in the hands of local religious authorities and distinguished this approach from that taken in water disputes.

This deference to religious authority appears in a case filed in 1919, but the measure of authority to which they deferred is one that colonial justices selected. In that case, lawyers invoked Section 295 of the Penal Code to settle a dispute over defilement of an idol in a village temple. Apparently, members of the Moothan caste (a division of the Śudrā caste) in Malabar had entered the sanctuary of a temple that was open to non-Brahmans. The prosecutor sought to prove that this entry was an act of pollution, because the low-caste group had been denied entry under local customary code. The Court disagreed with the petitioner's measure of customary code and decided not to convict the accused. It made the following statement:

> I accept the contention that the word "defile" cannot be confined to the idea of making dirty but must also be extended to ceremonial pollution, but it is certainly necessary to prove pollution. . . . The caste of the accused is not one of the polluting castes, vide Malabar Gazetteer, p.117. Neither is the act alleged, one confined by right to Brahmins as was the case in Sivakoti Swami (1) where a goldsmith touched the idol. It is simply presence in that part of a temple which

> is open to non-Brahmins but is alleged not to be open to the caste of the accused. In my opinion, this is not "defiling" within the meaning of S. 295 Indian Penal Code.[11]

In this case, the judge acknowledged the legitimacy of caste determinations of defilement but chose not to honor the determinations presented because they did not conform to the scale of defiling castes published in the gazetteer. The list in the gazetteer was the ecclesiastical measure the judge honored! So paradoxically, by relegating the decision to the gazetteer, the Court used a colonial rather than an ecclesiastical scale. While upholding casteism, they ultimately twisted it and revised it. The identity of the defiler, intentionally removed by government officials from well water disputes, was put into the symbolic center of decisions about the defilement of sacred icons and spaces. But this identity was subject to the particular measure that colonial officials selected.

In a case in 1924, the Court ruled in favor of a village custom that restricted specific castes from entering a local temple. In a village in the District of Chanda, there was an idol named Meghnath that was kept within an enclosed area of a temple. The idol was worshiped by all castes except the Mahars who were not allowed to enter the enclosure or touch the idol. They had to get a person of another caste to worship in the temple for them. On 17 September 1922, thirteen Mahars entered the enclosure of the temple in question, slaughtered a goat, and sprinkled its blood upon the idol. They also put *sindhur* (red powder) on the image and adorned it with flowers. For this they were convicted under Section 295 of the Indian Penal Code and sentenced to a fine of Rs 25 each. The sentences were upheld on appeal and the Court ruled:

> Now, there is no reasonable doubt that by long established custom Mahars have not been allowed to worship directly at the shrine and they must have known that to kill a goat there, to sprinkle its blood upon the idol or to touch the idol would, in this opinion of a large number of Kunbis and other castes, constitute defilement of the idol. In argument here the broad ground that Mahars as Śudrās have a right co-extensive with that possessed by all classes of Hindus to worship and touch all Hindu Gods is argued. This view in my

> opinion is far too broadly stated. When custom that has held for many centuries ordains that an untouchable, whose very touch is in the opinion of devout Hindus pollution, should not enter the enclosure surrounding the shrine of any Hindu God and when an untouchable with that knowledge deliberately enters and defiles it he has, in my opinion, committed an offence within the purview of S.295, Indian Penal Code.[12]

In this case, the justice did not restrict the notion of defilement to physical uncleanness but selectively extended it to include defilement of sacred and ritual objects. But like the case presented before it, the justice honored and then read into the power of "customs" on untouchability by selecting them according to his own discretion. This case, unlike the one before it, supported the restriction and the caste regime that produced it.

Returning to policies on well water, colonial authorities provoked a slow change in the way Hindus organized access to village-level supplies of drinking water. Through the interest in defining potability, government officials eventually connected different kinds of water bodies together under the same purview, extending their attention beyond wells to rivers and other freshwater sources. But within this purview, wells were, from the outset, treated differently. The Penal Code stated that a well could be regarded as a spring and a tank as a reservoir. However, a public spring could not mean a "continuous stream of water running along the bed of a river" ([Indian] Penal Code 1860:268). Therefore, court justices contended that cases involving the fouling of rivers could not be tried under Section 277 of the Penal Code. The defilement of a river could, however, be considered a nuisance under Sections 268 and 290 of the Penal Code,[13] which defined the nature of a public nuisance as "that act or process that causes common injury, danger or annoyance to the public or to the people in general who dwell or occupy property in the vicinity, or which must necessarily cause injury, obstruction, danger or annoyance to persons who may have occasion to use any public right."[14] The references to this distinction in the legal digests point out in an interesting way how judicial powers carved out a trajectory for advancing their control over rivers. Rivers, in contrast to village wells, were resources open to members of all caste groups, especially at their junctures with sacred places and complexes. Yet the interest in rivers rarely surfaced in legal settlements in the colonial period, appearing instead in policies on irrigation and the expansion of agriculture.

It is possible that the government's absence in dispute settlements regarding rivers was linked to the colonial government's absence in the politics of sacred spaces bordering sacred rivers. Riverfront areas—the ghāṭs described in chapters 4 and 5—are the direct points from which the public accesses river water resources. In colonial documents on Banaras, there is no mention of attempts by the government to control access to the river Gaṅgā or to meddle in use rights established by property owners and religious authorities in sacred spaces along the riverbank. Likewise, colonial documents on canal building in Haridwar do not mention attempts by officials to control sacred spaces along the river. Since in many places riverbanks are considered sacred, it is possible that government officials were leaving this control to the ecclesiastical domain and by doing so underplaying the issue of riverbank use rights. In Banaras, riverbank properties have been owned by kings, queens, saints, artists, and other prominent citizens and their upkeep supported through donations from patrons to local pilgrim priests. These properties have been controlled and maintained in a way that leaves room for the public to congregate directly on the ghāṭs as worshipers of Gaṅgā and other deities.

From Nuisance to Poison

As Galanter (1989:7) notes, the colonial legal code engendered a slow reworking of old concepts to make them fit new situations, concepts that were old in both British and Indian traditions. It did so by stretching slim precedents and confining uncomfortable ones, all under the pretext of merely following authority. The Indian Penal Code gave government offices legitimate rights over the definition of potability. Authorities used this precedent to expand the focus beyond wells to larger stagnant and flowing water sources. The shift toward rivers appears in prohibitions against certain usages of water bodies in the late nineteenth century but reveals a trajectory different from that developed to settle well water disputes.

This trajectory moved through laws that prohibited encroachments or obstructions that might impede transport and traffic in a waterway. The Shore Nuisance (Bombay and Kolaba) Act of 1853 gave the collector of land revenue of Bombay the authority to require the removal of any "nuisance, obstruction or encroachment" anywhere below the high-water mark in the Bombay Harbor, to facilitate safe navigation. The act sought to prohibit "nuisances" that might be created by dumping waste in the harbor. In other words, waste was seen as a potential irritant to naval transport and trade, not as a danger to the water itself. The Obstructions in Fairways Act of 1881 also empowered the central government to remove obstructions to navigation by regulating or prohibiting the disposal of rubbish in any fairway leading to a port (Manohar and Chitaley 1979b:36; Act XVI of 1881, cited in Mathur 1980:90). Both acts, while reflecting the interest in keeping waterways navigable for commodity trade, declined to define waste as a problem for use rights or for potability.

In other laws, however, there was some mention of protecting water bodies from contamination by foreign substances. The Oriental Gas Company Act of 1857 gave that company powers and facilities to light the city of Calcutta with gas. It included a provision requiring the Company to pay restitution or compensation in the event that it caused "fouling" of any water source (Mathur 1980:88–89; Manohar and Chitaley 1979b:167). In a subclause, the Northern India Canal and Drainage Act of 1873 also provided that any individual corrupting or fouling the water of a canal would be punished by imprisonment or fine (Mathur 1980:90; Act VIII of 1873).[15]

Moving closer to the interest in defining water quality, later acts showed a concern for the effects of fouling and contamination. The Indian Fisheries Act of 1897 prohibited the discharge of any poison, lime, or noxious material into a water source while intending to catch or destroy fish (Act IV of 1897; Manohar and Chitaley 1979a:446; Mathur 1980:91). A few years later, the Indian Ports Act of 1908 outlawed oil spills that polluted waterways, and after that the Indian Forest Act of 1927 moved to protect forest water sources from contamination (Act XVI of 1927). These laws show an evolving concern for preventing the contamination of water, but in actuality, they laid idle on the record. Though important as symbolic steps in the government's attempt to create hegemonic power, concerns about nuisances and contamination were never tested in the settlement of legal disputes.

While laws on nuisances and contamination in this period were never put to any use, they can be read as steps, albeit rather inactive ones, toward an interest in coopting rights to *define* water sources. River water policies were more closely linked to plans for irrigation and agricultural development (see chap. 5). Policies regarding the Gaṅgā surfaced in the making of the Yamunā and Gaṅgā canals in the late nineteenth century. At that time, colonial authorities aimed to divert a portion of the flow of these two great rivers to infertile land in the Doab. In a vein similar to the Obstructions in Fairways Act (1881), laws pertaining to canals prohibited obstructions and suggested that colonial authorities claimed control of channeled water. This assumption of authority to divert and channel river water fed into the controls officials later developed over other kinds of flows, such as flows of wastewater running through city sewers and drains.

In the twentieth century, judicial officials replaced the word *nuisance* with the notion of poison and brought water resources back into a more critical focus. The trajectory from the Indian Fisheries Act of 1897 to the Indian Forest Act of 1927 and later to the Factories Act of 1948 shows this shift in terminology and focus. The Fisheries Act stated a concern for the destruction of fish caused by the dumping of "poison, lime or noxious material" into water. Following this, the Indian Forest Act of 1927 made acts that "poison water of a forest area" offenses before the law (Mathur 1980:91–92). In 1948, the Factories Act took up the issue of industrial effluent and ordered its proper treatment and disposal.[16] While generally referring to large water bodies such as lakes, rivers, and seas, this legislation tended to emphasize the manner of causing pollution rather than the identities of polluters (Mathur 1980:93). This is a trajectory that started with well water disputes. But unlike well water disputes, these laws avoided attention to the consequences polluting practices could have on water quality. That focus did not resurface until the postcolonial period, but at that time for different ideological purposes.

Semantic shifts in the terms *defilement*, *fouling*, and *contamination* indicate that colonial officials were trying to create policies that would legitimize uses of water resources in new ways. These semantic shifts also tell us that colonial agents worked against very different kinds of customary rules on water resource use and realized that one way to undermine those rules was to discredit them through legal discourse. In the cases of rules on well water use, colonial justices discredited them by ruling on potability rather than on caste privilege or use rights. Potability was expressed through the terms *defilement* and *fouling*.

Given the paucity of data in this field, it is difficult to say what happened from the early 1900s until the end of the colonial period. Information on other cases involving well water disputes is simply not available in the Supreme Court library or in the Supreme Court digests. It is likely that a lag in legal activity did in actuality exist; mention of it has been made by contemporary lawyers and justices with respect to other judicial matters. For instance, it was not until 1987 that Supreme Court Justice B. N. Bhagwati threw out the British rule on strict liability to create a new rule on absolute liability to deal with industrial accidents affecting impoverished citizens. Enunciated in a British case a hundred years earlier, the strict liability rule had linked liability to an individual who brings onto his or her land something that is likely to do mischief or escape. Under this rule, an individual could be held liable for the damage incurred by the escape of this substance. In the postcolonial context, however, the rule provided plenty of loopholes for industries producing hazardous products. Moreover, the older rule did not apply when the escape of a substance was due to an act of God, an act of a stranger, or the default of a person (Mehta n.d.:10). In the aftermath of the Bhopal gas leak and in the context of a case on a gas leak from a Delhi industry, Justice Bhagwati created a new judicial norm of "absolute liability," requiring that a hazardous and inherently dangerous industry pay compensation for damages incurred to others. He stated:

> This rule, evolved in the 19th century at a time when all these developments of science and technology had not taken place, cannot afford any guidance in evolving any standard of liability consistent with the constitutional norms and the needs of the present day economy and social structure. We need not feel inhibited by this rule which was evolved in this context of totally different kind of economy. Law has to grow in order to satisfy the needs of fast changing society and keep abreast with the economic developments taking

> place in the country. As new situations arise the law has to be evolved in order to meet the challenge of such new situations. Law cannot afford to remain static. We have to evolve new principles and lay down new norms, which would adequately deal with the new problems, which arise in a highly industrialized economy. We cannot allow our judicial thinking to be constricted by reference to the law as it prevails in England or for that matter in any other foreign country. We no longer need the crutches of a foreign legal order. We are certainly prepared to receive light from whatever source it comes but we have to build up our own jurisprudence, and we cannot countenance an argument that merely because the new law does not recognize the rule of strict and absolute liability in cases of hazardous or dangerous liability or the rule as laid down in Rylands V. Fletcher as is developed in England recognizes certain limitations and responsibilities. We in India cannot hold our hands back and I venture to evolve a new principle of liability which English courts have not done.[17]

I have heard many other stories about instances in which lawyers in the 1980s had to actively encourage justices to throw out old and tired colonial precedents and reconstruct the legal landscape according to contemporary conditions (see chap. 7). However, other issues were at stake in the case of well water disputes. Apart from the fit of a particular law or rule to a particular period of government and commerce, semantic shifts in laws and rules also say something about the particular nature of the resource in question. I have suggested that the terms *defilement* and *fouling* applied to the potability of well water were eventually laid to rest and that later the notions of nuisance, poison, and contamination were applied to waterways, canals, harbors, and forest water resources. The latter terms, by contrast, were less significant to actual dispute resolution because they were rarely invoked in cases on water resources. In that sense, they meant very little to the accomplishment of hegemonic power. But there is another important point to be made. These shifts suggest that while village-level uses of well water were contested, uses of larger transregional water resources were not disputed at local levels. Rivers were of a distinctly different customary and legal character; though sacred to Hindus, they were not subject to control by high-caste elites who could attempt to restrict access and use rights. Thus they were not the focus of a colonial legal doctrine that sought to undermine caste-based codes, at least not until other considerations about river water use had assumed a more prominent position in economic development.

CHAPTER 7

Pollution and the Emergence of Judicial Activism

There are three basic commitments which the Supreme Court has developed in the last few years with a view to promoting human rights. One is the commitment to participative justice, the other is commitment against arbitrariness and third is the commitment to just standards of procedure.

—*Justice P. N. Bhagwati, Supreme Court of India, cited in Mehta n.d.:6–9*

A CONCERN WITH THE EFFECTS that polluting practices have on water quality did not reappear in an explicit way until the 1980s when government officials, legal authorities, and critical citizens began to engage in debates about uses and abuses of river water. These specific debates were far removed from those outlined in disputes over well water in the colonial period. Yet like colonial justices, government officials of this period claimed concern for water quality, but they did so by reversing their arguments. In the colonial period, justices had argued that individuals were not defiling or fouling well water, while in the postcolonial period they had to admit that individuals were able to pollute water resources. The difference was that the polluters and the effects had changed their look. Colonial laws were set up to produce a sociological effect—to define water in a way that would slowly permit access to disempowered caste groups and create instability in local power structures—and a hegemonic one—to control that resource in symbolic and, later, administrative ways. Individuals accused of polluting water resources in the colonial period were low- or outcaste group members. By the postcolonial period, the accused had become the elites, industrialists, and the general public.

From the 1980s, postcolonial law moved farther in the direction of water quality evaluation as a way to justify the government's evolving control over uses of river water. But this was, again, a tricky intention since water quality data had to be used to prove that polluters (who are now by and large the elites) are not, in fact, polluting. This has become even more critical as the polluter-pays principle evolves in legal discourse. Government officials were drawing from environmental discourses generated in international conferences to shape postcolonial laws, leaving precedents set by colonial officials to the distant past. In the 1980s, government officials began to adopt methods for measuring water quality and promoted them as tools for evaluating the consequences of intensive uses of the river that accompanied industrialization, urbanization, and agricultural development. While the colonial concern for the potability of well water had been rather inchoate in substance and not substantiated by other claims, scientific or otherwise, postcolonial programs, by the late 1980s, were using scientific evaluations of water quality to measure potability and argue for the success of cleaning projects. Forming the mainspring of policy-making activity, water quality data were used to justify the need for waste management infrastructure and to evaluate its performance.

New interests in international methods of water quality measurement and standards for potability now underlie the statements made by many officials of government regulatory bodies. In fact, the chairman of the most powerful regulatory body, the Central Pollution Control Board, wrote in 1992 that colonial legislation was not applicable to the conditions of postindependence India where the degradation of the environment was accelerating hand in hand with industrial development (Central Pollution Control Board 1995a:v). He suggested that foreign models of law and policy would better suit contemporary conditions. Colonial laws were devoid of scientific measures and thus inadequate, while modern measures used in global environmental protection and conservation projects demonstrated a more usable rationale.

The first provisions in this "protect the environment" mode of law were made in the Forty-second Amendment to the Indian Constitution. Passed in 1977, the Forty-second Amendment responded to the Stockholm Declaration adopted by the International Conference on Human Environment in 1972. That declaration laid out the essential right to live in an environment of quality and confirmed the responsibility of each member of society to protect and improve the environment. In conformity with these objectives, the Forty-second Amendment inserted Article 48-A into the Directive Principles of State Policy in Chapter IV of the Constitution. This declared the state's responsibility to protect and improve the environment and safeguard the forests and wildlife of the country. Another provision, inserted in Article 51-A(g), stipulated the duty of every citizen to "protect and improve the natural environment including forests, lakes, rivers and wildlife and to have compassion for living creatures." Both imposed an obligation on the government and the courts to protect the environment for the people and the nation (G. Singh 1995:50).

Along with this amendment, the Indian legal system provided a few other sources of law for addressing water pollution problems. The Water Act of 1974 was the first to specifically set out a concern for environmental protection. Parliament adopted minor amendments to the act in 1978 and revised it in 1988 to conform to the provisions of the Environment (Protection) Act of 1986. Although the Constitution had determined that water would be a subject in the State List and would therefore fall under the purview of the state, the Water Act enacted under Article 252 (1) of the Constitution empowered the Union Government to legislate in the field of state control. That power was affirmed when all states in the Union approved the act (Rosencranz et al. 1991:153).

The administrative regulation under the Water (Prevention and Control of Pollution) Act of 1974 provided for the establishment of a Central Pollution Control Board and, under this, a board in each state. These boards were supposed to develop plans for the control and prevention of pollution. The central board was to plan and execute a national program for preventing pollution, carry out research and compile data, and advise the government on water and air pollution matters. The state boards were responsible for implementing the Water Act by inspecting industrial and wastewater treatment plants and by conducting research on water quality and sewage treatment methods

(Rosencranz et al. 1991:154; G. Singh 1995:71–72). Both boards were given the authority to lay down standards for water and air quality and for emissions and effluents from industry and other sources (Central Pollution Control Board 1993–94).

The concept of pollution described in the Water (Prevention and Control of Pollution) Act of 1974 paralleled colonial notions of fouling and defilement of well water in that it lacked any consideration of caste identity and impurity or any intention to account for an ecclesiastical realm. It was a purely secular reading of water use. The definition of contamination, on the other hand, took on an ecological character, something that colonial notions of well water potability lacked entirely. Water pollution in the Water Act was described as:

> such contamination of water or such alteration of the physical, chemical or biological properties of water or such discharge of any sewage or trade effluent or of any other liquid, gaseous, or solid substance into water (whether directly or indirectly) as may or is likely to create a nuisance or render such water harmful or injurious to public health or safety, or to domestic, industrial, or agricultural or other legitimate uses, or to the life and health of animals or plants or of aquatic organisms. (see Section 2(e); cited in Central Pollution Control Board 1995a:2)

The 1988 Amendment to the Water Act introduced section 33A to empower state boards to issue directions to any person, officer, or authority to close, prohibit, or regulate the operation or process of any industry and regulate the supply of water, electricity, or any other service to the offending plant (Rosencranz at al. 1991:154). Moreover, the boards were granted the authority to hold corporate officials and heads of government departments liable for pollution offenses. This significantly augmented the power of the Pollution Control Boards. Prior to 1988, they were allowed to direct orders against polluting companies or individuals only in cases when they polluted "by accident or other unforeseen act or event." The boards were not able to take action against government officials, nor were citizens permitted to bring an action against a polluter. The Citizens Suit Provision of the 1988 Amendment Act gave citizens the right to bring actions against polluters.

Critics of the implementation of environmental laws have pointed out that the statutory duties of the Pollution Control Boards rendered them ineffective against government bodies and large corporations. Singh et al. wrote:

> Unfortunately, these Boards have little success to justify the responsibility reposed in them. The Boards are virtually defunct. Any ordinary law reporter will bear testimony to the fact that most environment cases are filed by organizations or individuals in public interest. Hardly ever does one see a case initiated by a Pollution Control Board. In fact, the Board often gets clubbed as party respondent (for dereliction of duty) with the offending company. The overemphasis and total monopoly of power vested in these Boards is misplaced. This is also probably the most important reason for the failure of these Acts. (n.d.:2)

In 1986, Parliament passed the Environment (Protection) Act to confer power on the central government to take measures to protect and improve the quality of the environment and to prevent, control, and abate environmental pollution. This act outlined a more holistic notion of pollution than the earlier Water Act had. In its preliminary statements, the act defined *environment* as the interrelationship that exists among and between water, air, land, human beings, other living creatures, plants, microorganisms, and property. Environmental pollution meant the presence of any solid, liquid, or gaseous substance present in such concentration as may be, or tend to be, injurious to environment.[1] Here, pollution damages an interactive environmental system whose borders are now significantly expanded. Quite unlike the definition of pollution provided in the Water Act, this piece of legislation left out explicit reference to the injury that pollution might have on public health or safety. This reference was transferred to the definition for hazardous substance. The act defined *hazardous substance* as that which is liable to cause harm to human beings, other living creatures, plant, microorganism, property, or the environment (Central Pollution Control Board 1995c:214). This suggests that government officials saw environmental pollution as a process injuring the environment but not necessarily the people who live in it. Hazardous waste, on the other hand, could render more pernicious effects.

Under the act, the central government was authorized to set standards for the discharge of environmental pollutants from various sources. The government was also authorized to close, prohibit, or regulate any polluting industry not complying with emission or effluent standards. They could also seek punitive action through the regulation of electricity or water supply to the plant.

Other laws (such as public nuisance actions under Section 133 of the Criminal Procedure Code, which allowed for citizen grievance, or the common-law right of riparian owners to unpolluted water) could have been used in pollution cases (Rosencranz et al. 1991:151). But neither of these laws was used. It was not until the late 1980s that the Supreme Court of India, when ruling on the Ganga Pollution Cases (discussed later in this chapter), added force to these laws by hearing public interest writ petitions submitted under Article 32 of the Constitution. Articles 32 and 226 confer power on the Supreme Court and the High Courts to issue writs in the nature of habeas corpus, mandamus, certiorari, and prohibition. These articles can be used to show that river pollution implicates a public authority who has been vested with the responsibility to prevent pollution but is not executing its powers. In effect, these articles are emancipatory provisions that allow citizens to challenge the structures of domination from within.

The Ganga Pollution Cases

In 1985, M. C. Mehta, a lawyer and social activist, filed a writ petition in the Supreme Court charging that, despite the strides made in the legal code, government authorities had not taken effective steps to prevent environmental pollution. Mehta issued a writ of mandamus to restrain leather tanneries and the municipal corporation of Kanpur from disposing of industrial and domestic effluent in the river Gaṅgā.[2] The court bifurcated the petition into two parts. The first dealt with the tanneries of Kanpur and the second with the Municipal Corporation. Called Mehta I and Mehta II in legislative digests, they became the "Ganga Pollution Cases," the most significant water pollution litigation in the short history of Indian environmental law.[3]

In the original petition of 1985, Mehta requested the court to order the leather tanneries of the Jajmau district of Kanpur to stop discharging their untreated effluent into the river. He also claimed that the

Municipal Corporation of Kanpur was not undertaking treatment of domestic sewage. The petition named eighty-nine respondents, among them seventy-five tanneries of the Jajmau district of the city, the Union of India, the Chair of the Central Pollution Control Board, the Chair of the Uttar Pradesh Pollution Control Board, and the Indian Standards Institute (G. Singh 1995:88).

During the first case (Mehta I) heard in September 1987, the Court invoked the Water Act and the Environment (Protection) Act in response to the petitioner's charges. Court officials pointed out that the state pollution control boards were not preventing the discharge of effluent into the river (*All India Reporter* 1988:1037–48; Malik 1988:319; Rosencranz et al. 1991:176; Singh et al. n.d.:186). The Development Council for Leather and Leather Goods Industries reported that tanneries from the Jajmau district were discharging five million liters of effluent per day into the river. The Hindustan Chambers of Commerce informed the Court that six tanneries had already installed primary treatment plants, and fourteen were engaged in constructing them. Other tanneries, through the Jajmau Tanners Pollution Control Association, registered their intent to install effluent plants for primary treatment of their industrial waste. After hearing the reports and affidavits, the Court gave the tanneries six months to complete the installation of primary treatment plants. After this, any plants not complying were to be shut down by the Central Pollution Control Board, the Uttar Pradesh Pollution Control Board, and the District Magistrate of Kanpur (Malik 1988:109–12).

The Court heard the second half of Mehta's petition in January 1988. Directing its judgment toward municipalities and towns situated on the banks of the Gaṅgā, justices opened with a discussion of the legal provisions and duties of municipalities. Their duties included treatment and disposal of sewage, provision of safe drinking water and public sanitation facilities, and disposal of human waste and dead animals (*All India Reporter* 1988:1115–28; Rosencranz et al. 1991:183; Singh et al. n.d.:192–93). The Court heard the specific instances pertaining to Kanpur and used these conditions as a model for generalizing about rulings on other cities in the Gaṅgā basin.

In this hearing, the Uttar Pradesh Pollution Control Board described the measures taken by local bodies and the Board to prevent river pollution. Their affidavit stated that the Kanpur Municipal Corporation (Kanpur Nagar Nigam) had not submitted any proposal for sewage treatment works to the board. The board deflected responsibility for pollution prevention to offices under the Ganga Action Plan. Those offices, created and financially maintained by the central government, proposed to expand the sewage system with financial assistance from the Central Ganga Authority, the state government, the World Bank, and the Dutch government. The Central Ministry of Environment and Forests explained that works under the Ganga Action Plan were to be carried out by three bodies: the Uttar Pradesh Water Engineering Board (UP Jal Nigam), the Kanpur Water Works (Jal Sansthan), and the Indo-Dutch project. The UP Jal Nigam, the state engineering and technical agency, was to remodel the sewage pumping station at Jajmau, improve the sewage farm, tap existing wastewater drains, and divert sewage to a treatment plant. The Jal Sansthan, the local water authority, was responsible for cleaning the trunk sewers. The Indo-Dutch project component included plans to lay an industrial sewer, build an upflow anaerobic sludge blanket (UASB) sewage treatment plant, and improve deficiencies in the city's sanitary facilities. The Dutch government provided most of the funds for these projects.[4]

After hearing reports, the court focused blame on the Municipal Corporation of Kanpur for the advanced state of river pollution in the city. It stated:

> On the facts and in the circumstances of the case we are of the view that the petitioner is entitled to move this Court in order to enforce the statutory provisions which impose duties on the municipal authorities and the Boards constituted under the Water Act. We have already set out the relevant provisions of the statute which impose those duties on the authorities concerned. On account of their failure to obey the statutory duties for several years the water in the river Ganga at Kanpur has become so much polluted that it can no longer be used by the people either for drinking or for bathing. The Nagar Mahapalika [the Municipal Corporation, later called Nagar Nigam] of Kanpur has to bear the major responsibility for the pollution of the river near Kanpur. (see Pandit 1989:244–45; Rosencranz et al. 1991:186–87; Singh et al. n.d.:196–97)

After fixing blame, the court ordered the Municipal Corporation to increase the size of sewers in labor colonies and lay new ones, construct public latrines and urinals, and ensure that dead bodies are not thrown in the Gaṅgā. This order was to apply mutatis mutandis to all other municipalities that have jurisdiction over areas through which the Gaṅgā flows. However, by neglecting to mention the responsibilities and accountability expected of Ganga Action Plan agencies, the order, in effect, misconstrued the structure of pollution prevention administration outlined in the report of the UP State Pollution Control Board. While the state board's report clearly reflected a new form of centralized management, the Court did not acknowledge this system when it laid blame on the municipal corporation. Commenting on this case, Singh et al. suggested that a kind of scapegoating process was at work. They wrote:

> Whether the court would pass such favorable orders against larger industries is a doubt which lingers in view of the unsuccessful prosecutions against multinationals and bigger industries. It was all very well to take stringent action against small scale tanneries, dairies or against municipalities who in any case never do their work. (n.d.:176)

Judicial Activism: Ascertaining the Correct Facts

Under the direction of the Central Ganga Authority, the UP Jal Nigam went about the task of renovating and constructing the sewage management infrastructure of Kanpur. Meanwhile, M. C. Mehta continued to press the Court to investigate the level of compliance achieved by the tanneries. In March 1990, the Supreme Court appointed an independent committee to visit Kanpur and "ascertain the correct facts" about compliance with its earlier orders (Malik and Malik 1991:181). The committee made its inspection and found that several tanneries had not taken any steps to install primary treatment plants. Some had installed the facility but were not operating it regularly, while others had treatment systems that did not work.

Before taking punitive action against the erring industries, the Court granted those with treatment facilities an additional six weeks to fix the defects and get their systems operating. Within this six-week period, the State Pollution Control Board inspected the tanneries and issued a certificate to those complying with the Court order. Those tanneries that had made no effort to install treatment plants were directed to close down until they had made the necessary working arrangements (Malik and Malik 1992:640–41, 1993:435).

By September 1991, the Court's directions paralleled the administrative steps taken in GAP projects in Kanpur. At that time, the Court enforced a scheme to collect contributions from each tannery to set up an additional secondary treatment plant (Malik and Malik 1992:635–41). The remaining funds for the construction of this plant were to come from GAP sources. Assisting GAP agencies, the Court directed the district magistrate and the Uttar Pradesh Pollution Control Board to close tanneries that had not paid their full contribution toward the secondary treatment plant. Forty-nine other tanneries were also closed because they had not set up their own primary treatment plants.

From 1992 through 1995, Mehta expanded his original petition to cover over 5,000 industries and 300 towns in the Gaṅgā basin. The Court set aside every Friday to review his petition along with other petitions that dealt with pollution matters across the Indian Union. In the meantime, the issue of the Kanpur tanneries faded into the background as other more pressing difficulties with belligerent industries took the Court's attention. Since industries were still reluctant to fully comply with Court orders to set up effluent treatment plants and treat their effluent to the standards prescribed by the central board, many Friday sessions were spent reinvestigating cases and imposing fines on or ordering the closure of negligent industries. In its biggest sweep, the apex court fined 191 industries in a single day. Over several years, it closed more than 500 industries in the Gaṅgā basin for failing to set up effluent treatment plants.[5] While following industrial compliance closely, the Supreme Court neglected to revisit the Municipal Corporations until 1993. In October, the Court directed the National Environmental Engineering Research Institute (NEERI) to supervise the municipal corporations in their installation of sewage treatment plants. NEERI found that of seventy-nine municipal authorities only nineteen had responded to the notices issued by the board. Many urban municipal corporations in the Gaṅgā basin had argued that they were not able to fully implement the orders of the Court due to lack of funds. Even those municipal corporations in the cities covered under the first phase of GAP could not demonstrate that they had actively participated in the works undertaken by the central government. Recognizing the marginal role played by urban institutions in these cities, the Court called upon NEERI to continue to monitor GAP activities in various municipalities.

In August 1994, NEERI's scrutiny turned toward officials of the central government, who, they claimed, were not fully executing work under the Ganga Action Plan. To debate this report, the Court called together the director of NEERI and the director of the Ganga Action Plan. The director of GAP provided reasons for the delay in completing projects, citing problems in collection and processing of data and electricity failure. These reasons did not suffice in the Court's view. Finding fault with the GAP administration, the Court decided to suspend any additional use of funds for the second phase of GAP until all works under the first phase were completed. This put a significant damper on GAP activities since phase two was slated to begin in 1995. Surprisingly, this order, issued in October 1994, remained in effect until 1996 because the discrepancies in evaluation reports between NEERI and the Ministry of Environment and Forests had not been resolved.[6]

After taking action against the Ministry of Environment and Forests, Mehta urged the Court to consider the role of the Central and State Pollution Control Boards, charging that the State Pollution Control Board was not monitoring industries according to the standards prescribed by the central board. Alleging that favoritism guided the state board's monitoring of industrial effluent, the team argued that the state board's inspection reports were "made mechanically without indicating the quantity of effluent discharged by the individual industries."[7] Mehta's team claimed that in place of actual measurements of effluent discharge, the state board was making reports "by mere assessment." Hearing this, the Court directed the Central Pollution Control Board to evaluate the state board's measurements, analyses, and reports and to indicate their findings to the Court.

On 20 October 1995, the Court recalled the issue of the UP state board's inspection competence. After hearing statements from the state board, the central board, and the petitioners, the Court made the following order:

> We do not wish to proceed any further with this matter. We drop the proceedings/action initiated against the Uttar Pradesh (UP) Pollu-

> tion Control Board. We direct that so far as the functioning of the UP Pollution Control Board in controlling pollution and assisting this Court is concerned, it shall function under the supervision of the Central Pollution Control Board.
>
> So far as the functioning of the two Boards under the Water and Air Prevention Control of Pollution Act, 1974 is concerned, the UP Pollution Control Board shall be bound by all the directions issued by the Central Pollution Control Board.
>
> The primary responsibility in pollution control matters over the industries in the State of UP shall be of the Central Pollution Control Board. All the reports shall be counter-signed by the Central Pollution Control Board. In case the Central Pollution Control Board feels that a second check is necessary a team of the Central Pollution Control Board shall go and make a second check in respect of the individual industries. This arrangement shall continue till further orders.[8]

The state board was caught derelict in its duty to hold industries to environmental regulations, Supreme Court orders, and notifications issued by the Ministry of Environment and Forests. While the Court treaded lightly by dropping the actions against senior officials of the Board, it enforced disciplinary measures by calling on the central board to supervise. This put considerable pressure on the central board to act as an authority over the state board, but relieved the central board of the responsibility of making the decision to police the state board on its own. In this intervention, the Court averted a potential conflict of political interests between the state government and the central government who appoint the chairmen of the state board and the central board respectively.

In late 1995, the apex court encouraged M. C. Mehta to conclude the Ganga Pollution Cases by recommending strategies for remediation. At that juncture, Mehta was concerned with issues of accountability and corruption and sought a way to centralize the work of treatment plant construction. The idea was to require that all accountability rest in one place. Ideally, he thought this would reduce the number of levels at which corruption could occur. In November 1995, the Court agreed to present this proposal to the National River Conservation Directorate, the new directorate created within the Ministry of Environment and Forests to encompass the Ganga Action Plan and other river action plans. The ministry rejected the proposal, arguing that a central body to supervise treatment plant construction would undermine the new matching fund agreement for phase two of GAP. In this agreement, 50 percent of the costs of new plants are to come from state resources. The central government pointed toward a new constitutional amendment, the Seventy-fourth Amendment, which advocates decentralization in administrative and fiscal matters. But Mehta remained skeptical. In his view, this interest in centralized authority without centralized accountability would only continue to allow trickle-down corruption. In early 1996, the justice hearing the Ganga Pollution Cases retired, and the case took a back seat to others in the Supreme Court ledgers. Though still pending, by 1999 the case had no real direction from the bench, forcing Mehta and his legal team to turn to other strategies.

For water quality, these developments in law and the Supreme Court's activism have meant two things. First, by the late twentieth century, the quality of Gaṅgā water became inextricably tied to a maze of institutions, bureaucracies, and court orders. Moreover, since it was no longer possible to argue that water quality could be maintained at levels set by the World Health Organization in 1984, the game of manipulation began to assume serious proportions.[9] The real purpose of these manipulations has been to apportion blame to others and avoid the label *polluter.* The maintenance of city sewage, the provisioning of city water supply, and the functioning of local, state, and central bureaucracies are all tied up with the assessment of Gaṅgā's water quality.

PART 4

CHAPTER 8

The Institutions of Wastewater Management

THE CONCERN FOR WATER QUALITY stood as the original justification for government policies and programs that began to assume lives of their own. As the government expanded its activity in the area of urban wastewater management, it used policies on water quality to nurture the growth of these systems. The stated goal was to restore river water to standards of potability and bathing set by the World Health Organization. This focus on the restoration of acceptable levels of water quality developed into a number of technological applications and was absorbed into the wider scientific discourses of engineering, hydrology, river ecology, urban geography, and city planning.

In cities along the river Gaṅgā, engineers were building modern systems of wastewater treatment and disposal over older drainage systems. The older systems had relied upon gravity to dispose of wastewater, while the newer systems were circumventing these gravity paths and moving wastewater with electric power. However, this overlay of new system upon old, while meeting some of the needs posed by increasing wastewater flows, created a new set of infrastructural problems that officials had to find ways to deal with. Today, wastewater management agencies in urban centers in Uttar Pradesh confront two related challenges: to treat and dispose of wastewater and to rectify problems created by the overlapping of gravity and electrically based systems. Both of these challenges find their origin in the shift from dry conservancy to water-carriage methods of waste disposal. About Australia's relation to this particularly European method, Beder has written:

> The sewerage engineering paradigm is firstly based on water-carriage technology. The struggle between water-carriage technology and dry conservancy methods of dealing with sewage took place in the nineteenth century. Water-carriage technology triumphed on the basis of theories, beliefs and values which were held at that time (Tarr 1984; Beder 1989). The advocates of both water carriage and dry conservancy methods relied on scientific theories that are largely discredited today. The water-carriage lobby argued that organic wastes had to be removed from places of habitation as soon as possible because if they were given time to putrefy or decompose they would give rise to 'miasmas' or disease producing gases which were responsible for the spread of diseases such as typhoid. Water-carriage enabled these wastes to be whisked away immediately whereas dry conservancy methods required that the wastes be stored about the premises (Sewage and Health Board,1875, p6; Tarr 1984). (1993a:82)

This chapter presents a thick description of wastewater management practices in Banaras and Kanpur, bringing attention to bear on the material configurations of wastescapes and the transformations in their flows over time.

"The Benaras Stinks"

In sixteenth-century Banaras, Moghul rulers laid the first drains to carry storm water away from the city and into the river Gaṅgā (Sankat Mochan Foundation 1992). However, it was not until Duncan, the first British Resident of Banaras, set out to improve public sanitation that a disdain for waste appears in the administrative record. He expressed disgust for the lack of public sanitation when he diverted fees and fines collected in the courts of Banaras to construct public latrines and waste receptacles. The people of Banaras responded by interpreting these measures as government intervention and opposing them.[1] Frustrated, Duncan complained to the government in October 1792 that "the inhabitants were so invincibly bent on adhering to their old practices, however prejudicial to their own healths as well as to the general cleanliness of the place."[2] Determined to act further, Duncan proposed that a drain be laid from the garden of Beni Ram Pandit to an outlet on the Varuṇā River. This would carry off stagnant water collecting in city streets and walkways *(galis)* and eradicate conditions in which filth and disease might breed (Government of India 1965:280–85; Narain 1959:182–83).

Despite Duncan's efforts, the plan was not seriously considered until the superintending engineer of Allahabad resurrected the case a hundred years later. In 1880, superintending engineer Fitzjames wrote the *Preliminary Report on the Sewerage and Water Supply of the City of Benaras* in which he outlined a large-scale sewage and storm water drainage plan for the city. In this report, he made comments about conditions of city "filth" to give legitimacy to his pursuits:

> When I state that it may be said to be without drainage of any kind, that its subsoil is saturated to a depth of several feet with the filth and abominations of centuries, that every well in the city is contaminated by percolation from the offensive soil, and that in spite of all these evils it is a healthy city, one wonders how this can be. . . . The Benaras stinks are already gaining a world-wide celebrity, and ere long those of Cologne will be forgotten. These stinks only too plainly and emphatically tell us how much drainage is wanted in Benaras and there cannot be much doubt that if this city is not very thoroughly drained and improved, as much as sanitary science will allow, an outbreak of some epidemic like the plague will devastate the city, and cause heavy loss to Government. (1880:9)

He continued by linking the problem of filth to a denunciation of the "natives":

> To realize the filthy state of Benaras is quite beyond the power of anyone who has not visited it. The subsoil, when opened out, gives out a most sickening offensive stench. I firmly believe that if the proposed drainage works are carried out, we shall have to use deodorizers very freely, and in some parts of the city, I shall not be surprised to hear of the work-people being driven out of the trenches by the stinks, though we all know natives can stand a good deal in that way. (1880:11)

Like other British officials, Fitzjames criticized the unsanitary conditions of Banaras, mustering scientific rationalism in an assault against Hinduism and local culture.[3] Moreover, his position paralleled the colonial government's interest in defining the potability of wells; he did not in an immediate way attempt to usurp physical control of urban space or water resources. Rather, he and other colonial officials slowly built up the codes to legitimate more expansive administrative practices. Omitting any reference to religious sentiments toward the Gaṅgā, he drew instead from discussions on waste management current in England and America. Fitzjames, like colonial justices, separated the civil from the ecclesiastical domain.

Fitzjames designed a main sewer line to run through the heart of the city, from Durgā kuṅḍ temple in the south (and upstream) to the mouth of the Varuṇā River just before its confluence with the Gaṅgā. Lying in a natural valley, the line was designed to carry sewage by gravity and become a catchment artery for other lines that the municipality would connect to it over time. Ideally, the system would separate storm water from sewage by diverting storm water to open drains along the streets and wastewater from households to the underground trunk line. To manage the heavy load of storm water during the monsoon, Fitzjames stressed the importance of keeping the two flows separate.

The problem was that the sewer had to be cut into the earth at a downward diagonal. It was to start at the upstream end of town just below the surface of the street and end at the downstream end of the town at a depth of thirty feet below street level. This meant that wastewater had to be pumped up out of this hole by electrical pumps and then directed into the flow of the Varuṇā River. This would divert wastewater away from the city and the bathing ghāṭs without treating or cleaning it. Unique for the history of the city, it combined the use of gravity and the use of electricity to move and circulate wastewater. Over time, however, this approach created a series of new problems for the city's residents.

In 1886, the Kashi Ganga Prasadini Sabha (the municipal board) approved Fitzjames's plan, and by 1892 the actual projects were completed (Arnold 1989:265; Government of India 1965:283; Varady 1989:240). The board also upgraded the water supply system to introduce more water volume into wastewater flows. By 1898, the board was providing drinking water through two main pipelines to 6,000 houses and 400 standposts and wall fountains (Government of India 1965:284). By 1956, domestic tap connections had increased to 20,393 (Kumar 1988:107). Meanwhile, sewer lines were expanded to meet the requirements brought on by an increased water supply. From 1913 to 1917, the city municipal board connected the Orderly Bazaar sewer and a sewer running underneath the ghāṭs to this main trunk line. The sewer under the ghāṭs diverted sewage away from the main bathing area around Daśāśvamedha. At that time the total city discharge of wastewater was about five million gallons per day, and the volume of river water was 1,200 times greater than the volume of the city's wastewater (Hammad 1992:27). Given this ratio, the river could regenerate itself fairly quickly.

Over the next half century, this infrastructure deteriorated. The Orderly Bazaar sewer washed away in the 1943 floods and, after that, other breaches occurred in the main line. The monsoon of 1948 washed away the ghāṭ sewer, and a cast iron pipe was laid in its place. In the 1950s and 1960s, repairs were made to the ghāṭ sewers, while additional branch sewers were connected to the main trunk line. Apart from these piecemeal repairs, however, government officials remained financially ill-equipped to chart out larger proposals for a more comprehensive pollution prevention scheme (Hammad 1992).

Reports show that by the 1970s, human waste from settlements developing on the upstream side of Banaras and industrial effluent from the city's largest industry, the Diesel Locomotive Works, were flowing directly into the Assī Nālā. This transformed the Assī, a creek that once formed the southern border of the sacred city, into an open drain for wastewater. Five sewage pumping stations installed along the ghāṭs in the early 1970s pumped the sewage flowing by gravity toward the Gaṅgā back to the main trunk line. But after a short time, these pumps ceased to work and lay dysfunctional until the latter half of the 1980s. At that time, the government initiated restoration under a new environmental program, the Ganga Action Plan (Uttar Pradesh Jal Nigam 1991).

The Ganga Action Plan

By 1981, debates in the Lok Sabha, the People's House of the Indian Parliament, began to produce a series of official statements stipulating the use of the Gaṅgā as a national resource. When Rajiv Gandhi announced the Ganga Action Plan in 1984, he drew legitimacy from the Constitutional Amendments created in the 1970s. Over the next decade, discussions of waste management in Parliament developed three themes: the definition of official responsibility for pollution abatement, the development of waste management infrastructure, and the measurement of pollution. The measurement of pollution provided data that was used to justify the need for waste management projects under the Ganga Action Plan and prove the long-term efficacy of these programs.

When Rajiv Gandhi created the Ganga Project Directorate as the authoritative body to oversee the Ganga Action Plan, he envisioned a modern bureaucracy, one that would respond to the needs for national development and the demands of the global market at the same time. The plan approved by the government in April 1985 pursued two objectives: to reduce the pollution load in the Gaṅgā and to establish sewage treatment systems in twenty-five Class I cities bordering the river (see Alley 1992; Uttar Pradesh Jal Nigam 1991). Shortly after the creation of this authority, the minister of environment and forests began a vigorous campaign for foreign investment, one that pulled in generous grants from the Netherlands and large loans from the World Bank and development banks in the United States, England, France, and Japan. While this foreign capital was solicited, the Planning Commission proposed a 250 crore (2.5 billion rupee) expenditure to complete the first phase.[4] Soon after, parliamentarians began to announce various schemes and their costs in the Lok Sabha. Assuming a frantic pace, these announcements began to elude a clear sense of program costs.[5]

One of the first schemes promoted by the government with NGO support focused on the cultural gentrification of Maṇikarṇikā ghāṭ, one of two sacred places in the city where Hindu cremation rites occur. This site was selected because officials saw that religious practices were having harmful impacts on river water quality. Yet, the project focused solely on the physical structure of the ghāṭs rather than on river cleaning or the modification or elimination of cultural practices. With funds provided by GAP and a nongovernmental organization called the Indian National Trust for Art and Cultural Heritage, the cremation ghāṭ was repaired, and facilities such as steps and railings were added. However, because officials failed to take into account the interests of those who controlled the sacred space of Maṇikarṇikā, the project did not result in a long-lasting solution. After a short time, workers at the cremation ground and other residents in the area resumed control of the gentrified areas and slowly reformed the "improvements" to meet their own work and living needs. Today, the ghāṭ looks much like it did before the restoration project began, and the cement structures on which the funeral pyres were built are again in need of repair.

In 1987, the environment minister created a police force to patrol the ghāṭs of Banaras and prohibit the following activities: defecation along the bank, dumping of construction debris, waste, and animal carcasses, and the indiscriminate mooring of boats. The Lok Sabha Debates also publicized a scheme to utilize an endangered species of turtle to, as the government office put it, "maintain the ecological balance in the river." To "clean" the Gaṅgā, they released the turtles into reaches of the river to consume the flesh of floating corpses.[6]

The older system of drainage in Banaras worked upon gravity and the flow of the river, while the modern system sought to use electrical pumping and mechanical aeration to dispose of and treat effluent. Today in many locations electrical pumping works against gravity by carrying wastewater away from the places it flows because of the topography of terraformed and built environments. Where the two systems (the one driven by gravity and the other by electricity) meet, the gravest urban problems arise. This intersection between old and new systems of sewage management is most evident along the ghāṭs. By gravity, wastewater flows through open and underground drains down the slope toward the Gaṅgā. These flows pass over and under the homes and businesses that have crowded older paths of drainage. When wastewater reaches the river's edge, the two systems intersect.

I first realized in a profound way how urban topography shapes wastewater flows when I was caught inside a home on Daśāśvamedha ghāṭ during a torrential downpour. When I looked out the window, I saw a vicious stream running from the city past the house and toward the river. If I had attempted to make my way back home at that time, the rushing water would have swept me away. This drainage mixed storm water with the wastewater that normally flows through open drains toward the Gaṅgā. In drier times, wastewater is more concentrated with domestic sewage but continues to flow along the same drainage paths toward the Gaṅgā, making the riverbank a sieve for urban effluent.

Wastewater flows are also complicated by the accumulation of silt and solid waste in the sewer lines. The silt and solids flow along with the wastewater as far as gravity takes them and then build up in the interstices of the underground tunnels. Siltation can also occur when a sewer line or open drain is blocked or closed for other reasons. To avoid a buildup of gaseous pressure, the local municipality removes as much silt as it can from manhole points along the main trunk lines; however, many areas are not accessible with the technology that they use. The more serious siltation problems tend to occur at the points where the drains terminate under the ghāṭ floor near the river, which is where wastewater is supposed to be trapped and pumped back into the city to meet the trunk line that runs downstream. To pump these flows, Jal Nigam technicians close the gates under the ghāṭ floors to prevent wastewater from flowing directly into the river. However, when electricity is not available to activate the pumping, wastewater builds up quickly in the sewer lines under the ghāṭs. At these times, silt and solid waste accumulate at underground points along the ghāṭ, produce noxious gases, and cause ruptures in the ghāṭ floor and the streets. One day, without warning, gas and solid waste broke through the stone floor on Daśāśvamedha ghāṭ and shot up into the air like an erupting volcano. To avoid these events and alleviate the pressure from gaseous buildup, Jal Nigam officials open the gates under the ghāṭ floor and bypass sewage directly into the river. They do this at all the pumping stations along the riverbank when the power supply is cut off.[7] At these times, wastewater flows into the Gaṅgā at spots where pilgrims bathe and perform ritual ablutions (see fig. 9).

Most of the sewage of the city is supposed to be pumped mechanically to the main pumping station at Konia and then diverted to an activated sludge treatment plant at Dinapur. However, the collection chamber at Konia often overflows. During the monsoon when runoff from the open nālās also enters the sewer lines, this overloading increases exponentially. Again, at those times, Jal Nigam technicians divert wastewater overflows into the bypass drain near that pumping station so that it runs directly into the Gaṅgā. When the wastewater is pumped in its diluted form into the plant, the food-to-microorganism ratio drops and the efficiency of the treatment plant is

Fig. 9. Escape valve for the pumping station on Rājendra Prasād ghāṭ (adjacent to Daśāśvamedha ghāṭ)

reduced. At these times, the activated sludge process cannot reduce biological oxygen demand and suspended solids in wastewater to the required levels.[8] So wastewater is either directly diverted to the Gaṅgā or treated in an ineffectual way before it is directed back to the river.

Official Environmentalism and the Success of Disjunctures

In 1988 when the Supreme Court accused the central, state, and local authorities for failing to implement environmental regulations, they were pointing to the disjunctures between policy and practice that have laid the groundwork for the double-sided nature of official environmentalism today. Although they are called failures by court justices, these disjunctures have to be seen in terms of their advantages, as exercised by individual bureaucrats and industrialists. The many disjunctures created by the contradictory legislative and institutional history of waste management may allow an official or an industrialist to achieve private goals in an otherwise complex institutional system. As Ferguson (1990) argues, the guise of failure does not tell us that something is inherently wrong with third world cultures, as if they are not able to play the game of capitalist development well enough. The image of failure hides other activities and operates for other priorities. Disjunctures help bureaucrats negotiate bureaucratic rules to generate personal successes out of national failures.

During a hearing of the Ganga Pollution cases in 1988, the Court lamented:

> It is unfortunate that although Parliament and the State Legislature have enacted the aforesaid laws imposing duties on the Central and State Boards and the municipalities for prevention and control of pollution of water, many of these provisions have just remained on paper without any adequate action being taken pursuant thereto. (*All India Reporter* 1988:1116; quoted in Rosencranz et al. 1991:183 and Singh et al. n.d.:193)

Complaints that things "on paper" are not carried out in ground realities can be heard today in casual conversation in any Indian city or village. Local politicians, religious leaders, and members of independent research bodies use the "on paper" idiom to blame official government and its various bureaucratic offices for fundamental failures. Journalists in the popular press often accuse government officials of creating and then benefiting from inefficient operations. As nongovernmental organizations are quick to point out, policies in government agencies move on one plane while official practices work on another.

The institutional complexity that acts as a backdrop to these disjunctures is somewhat clearer in Kanpur, because foreign donors have spent considerable time trying to understand that complexity and have produced numerous analyses of bureaucratic structure and rules and of disjunctures between institutions that are in some cases what foreign agencies and donors helped to create. In Kanpur, wastewater is generated by a population of four million residents and by 200 large and small leather tanneries, an ordinance factory, cotton factories, heavy chemical factories, food processing units, and paper and pulp industries (see Schjolden 2000). The Indo-Dutch project, a program fully funded by the Dutch government but executed by the government of India, installed two wastewater treatment plants with capacities to treat five and thirty-six million liters per day (mld). These plants employ a Dutch technology, the upflow anaerobic sludge blanket (UASB), to treat a mix of industrial and domestic wastewater. This method is different from the method employed in Banaras and represents the direct interests of the Dutch donor. Under the Indo-Dutch project, sewer lines and pumping stations have also been renovated, and a new industrial sewer was constructed to divert the wastewater from leather tanneries to the thirty-six-mld treatment plant. At the end of the first phase of the plan (1994), the Dutch government began to consider allocating an additional grant of one billion rupees to expand the sewerage system in South Kanpur.

In Kanpur, maneuvering by several agencies involved in wastewater management creates this institutional complexity. The Uttar Pradesh Jal Nigam positioned itself to become the darling of the central government's Ganga Action Plan and receive the lion's share of spin-offs from the construction of sewer lines and treatment plants.[9] Meanwhile, local institutions such as the Jal Sansthan and the municipal corporation were marginalized and grossly neglected in administrative and fiscal terms. This marginalization resulted from two processes. First, the central government, by assuming the role of constructing sewage treatment and disposal infrastructure, usurped without legal right the responsibilities held by these local bodies. Second, the Supreme Court pushed municipal corporations further into a marginal position when they used them as scapegoats for city waste problems (see chap. 7). Municipal corporations in urban centers along the Gaṅgā responded to the court's accusations by growing more distant from the work that the Court ordered them to do in 1988 and that the Ganga Project Directorate ultimately took up.

Recent contests aside, the deeper roots of this institutional complexity and its resulting disjunctures can be found in long-term state policy priorities and contradictions in legislative developments. Over the past three decades, policy initiatives for guiding urban municipal institutions have been dominated by a planned, rurally oriented, state-directed, top-down approach (Rosenthal 1970). This has disconnected the statutory institutions responsible for planning, delivering, and managing urban services from each other, from consumers and users, from access to private sector resources, and from community participation in service delivery.[10] Urban municipal authorities were divested of the power to run their own waste management institutions and use those tax revenues to run local schemes. Power and money remained with central and state officials who took no interest in raising tariffs as a way to fund wastewater management at the local level. As a result, urban institutions have remained supply-driven and depend on state grants and donor-funded projects to provide services to the public. In their underfunded state, they provoke widespread public distrust with their less than satisfactory water and sewer services. The public expresses this distrust by refusing to pay water and sewer bills. Poor cost recovery has been compounded by the fact that the public generally considers water, sewerage, solid waste, and sanitation facilities as social or welfare services that government should provide at little cost to the consumer.

The consolidation of administrative and fiscal power at the central and state levels also belies a contradictory legislative history. Under pressure from the World Bank in the 1970s, the government passed the Uttar Pradesh Water Supply and Sewerage Act to give responsibility for providing and managing urban waste and sewerage to the state agency, the Jal Nigam. Prior to the act, municipal corporations (then called the Local Self-Government Engineering Department of the state of Uttar Pradesh and the Water Works Department, or Jal Kal Vibhag) held authority over the management of these urban services. Before 1975, the Jal Kal Vibhag was under the administration of the municipal corporation. The city engineer (Nagar Abhiyanta) headed both the Water Works Department and the Local Self-Government Engineering Department and was appointed by the mayor. The UP Water Supply and Sewerage Act created the UP Jal Nigam as the state water engineering board and the Jal Sansthan as the local works authority responsible for water and sewerage service and maintenance. Moreover, it removed the position of city engineer and created a general manager of the Jal Sansthan, an official who had to be appointed by the state government and was accountable to that higher body. Another official position, that of the city administrator (Mukhya Nagar Adhikari), was created by the state government. This official was appointed by the state government to work with the mayor and oversee the municipal corporation.

Participatory democracy changed as a result of these legal and institutional shifts. Elections for the post of mayor, elections formerly held by popular suffrage, were put into the hands of corporators, the neighborhood-level representatives elected by the people. These changes brought about a diminution in the power of the mayor vis-à-vis other state-appointed officials. The policy of state appointment of the general manager of the Jal Sansthan and city administrator also meant that these local leaders could be changed as often as state-level officials were or needed them to be, according to the whims of politics. The policy of transferability prevented the general manager of the Jal Sansthan and the city administrator from becoming too powerful in a local area and kept control in the hands of the state government.[11] This diminution in the power of the mayor led to a diminution in the power of the municipality as a whole.

This centralization of government power has also shaped the operation and maintenance of sewage management infrastructure. After constructing wastewater treatment plants and conveyance sewers under the Ganga Action Plan, the UP Jal Nigam retained the responsibility for maintaining treatment plants and sewers in the cities covered by the Ganga Action Plan. This directly contradicted the 1975 Water

Supply and Sewerage Act, which stipulated that operation and maintenance must go to the Jal Sansthan. Since the inception of the Ganga Action Plan, the UP Jal Nigam, with state government power behind it, has claimed rights to operate and maintain the more costly sewage treatment plants, arguing that they are the only professionals with the technical expertise to do the job. But this claim is selective: in Kanpur, they take responsibility for only those plants that fall under the purview of Dutch-financed projects. Selecting these, they position themselves for access to monies allocated for the operation and maintenance of these facilities. Meanwhile, they hand over the less technical tasks of cleaning and maintaining the domestic sewerage system to the Jal Sansthan. These are tasks that are not likely to be funded under the Ganga Action Plan. All other sewerage needs go unattended as the Jal Nigam maintains control of only those facilities that are slated for continued assistance or that fall within the purview of larger, funded action plans.

While planning the fiscal requirements for the second phase, the Dutch government decided to link the allocation of grant money to policies and practices aimed at strengthening local institutions and promoting human resource development within them. To do this, they created an additional project under the name Institutional and Community Development Project (ICDP) (BMB et al. 1995a, 1995b). This formed the "soft wing" of the Dutch government's technological interests and was in keeping with the long-term understanding that projects undertaken with GAP funding or foreign grants would be turned over to the municipal corporation and the Jal Sansthan for operation and maintenance. This was a commitment that the Central Ganga Authority had endorsed but not built into its own resource allocation plans. In fact, the Central Authority has at various times explained to the Dutch government that their mandate does not include support for local urban institutions to administer operation and maintenance over the long term. Since the accountability of the Kanpur municipal corporation worsened considerably during the period of the Ganga Action Plan, the Dutch government has begun to look for some long-term institutional solutions to sewage problems. The Indo-Dutch project required that action plans for Kanpur be designed for the institutional unit despite the changes in personnel that often take place in high-level posts over time. The idea was to avoid the possibility that their plans would be dismissed if personnel were shifted. They asked the Ministry of Environment and Forest to recognize the need for restructuring civic bodies; the ministry agreed in theory but made little attempt to work in the integrated project.[12]

Unfortunately for Dutch officials, the objectives of the soft wing of the Indo-Dutch project have been slow to materialize. From 1996 through 1998, there was little money allocated to carry out the larger-scale institutional changes that were envisioned. Many fine surveys were conducted and research reports written by Dutch and other European consultants, but actual improvements in the Jal Sansthan's revenue-generating capacity remained negligible. After several interviews with the directors of ICDP, I have come to realize that there are two aims to this work. The one aim is tied to the infrastructural provisions needed to do such things as update and computerize property tax records, computerize billing for water and sewer services, clean sewer lines, make minor repairs, and maintain the waterworks system. The second aim is to transform the institutional organization of power, an organization whose deep structure changes, if at all, ever so slowly. The intent of the Indo-Dutch soft project was to strengthen the power of local bodies and take decision-making control out of the hands of state and central government offices. Dutch and European consultants intended to undermine the monopoly held by the Jal Nigam by constituting a "Project Preparation and Control Unit" (PPCU) consisting of the Jal Sansthan, the Nagar Nigam (the city municipality), and Indo-Dutch officials, in effect a decision-making unit at the city level. Originally, the idea was to give powers to decide sewage management issues in the town to the PPCU. The PPCU would then decide on the contractor to use in sewage engineering projects. This would reduce the Jal Nigam to one of several bidders for the work.

This second aim has been nearly impossible to achieve. The central government has shown no intention to change the structure through which power and money flow in this field. When the central government receives a foreign grant, it is administered through the Ministry of Environment and Forests at the central government and then directed through the government of the state in which a project is situated. The state government routes the funding through the State Ministry of Urban Development, which then elects the Jal Nigam to handle construction, engineering, and technical tasks. To control construction is to control the flow of money in this field of development, and the Jal Nigam is a structural ally for state officials who are responsible for translating foreign grants and loans into infrastructural projects at local levels. The Jal Nigam is a state body, so all urban units of this agency are directly tied to the state government. To cut the Jal Nigam away from this monopoly role would require political will at all these levels. Since political will runs along other lines, toward the central government rather than toward the periphery and to local municipalities, change has been minimal.[13]

One might expect that the agencies assuming control over the administration and construction of waste management systems would have also appropriated the rights to collect the revenue needed to run the projects. But this has not been the case, again, because a contradictory legislative process has created a series of disjunctures. For example, while the Jal Nigam became a key player in the construction as well as operation and maintenance of sewage treatment plants in urban centers, the Jal Sansthan continued to hold the statutory power to levy tariffs or charges for water and sewerage services. While the Jal Nigam has had to pay for maintenance tasks, the Jal Sansthan has collected the revenues for water and sewer services. The theory of the law was that the Jal Sansthan would hold the power to levy taxes for the maintenance work it would carry out. But when the Jal Nigam usurped operation and some maintenance tasks under the Ganga Action Plan, revenue collection was disconnected from the maintenance of infrastructure.

Quite curiously, this disjuncture has not led to a greediness within the Jal Sansthan to make profits from the taxes they are empowered to collect, since the maintenance of this infrastructure is provided by another agency. The Kanpur Jal Sansthan has been so marginalized by the government and the court that it has sunk into a real condition of atrophy. In this marginal position, municipalities assume the unattractive public image of a provider of poor service. This means they encounter great difficulties collecting service charges and in the process lose interest in collecting for the local government. Instead, residents allege that officials make small businesses out of what is left of tax collection routes. This makes the whole idea of controlling tax collection rather unappealing given other opportunities to work with state and central grants.

There is one recourse Indo-Dutch officials appear to see, and this is empowerment through the more permanent officials who work within the Nagar Nigam and the Jal Sansthan. All project approvals are made through higher-ranking officials or Indian Administrative Service (IAS) officers sitting in the positions of city administrator (Mukhya Nagar Adhikari), district magistrate, and commissioner. The understanding is that these officials will be shifted from their positions within a year or so of taking office. The more permanent, lower-level staff, though largely powerless as individuals, give some hope for change at the local level because they outlast the transfers of high officials and give continuity to the functioning of these local bureaucracies. There are indications that some reforms can be built up from coalitions of lower-level staff and foreign consultants.

This said, it would be important to turn for a moment to the water supply situation in Kanpur to make sure that concerns for wastewater do not completely overshadow concerns for water supply issues or that their interconnectedness is not lost. The relation between water supply in a city and wastewater output is closely linked, so the one system must be kept in perspective when discussing the other. In Kanpur, while city sewage infrastructure received considerable attention under the Ganga Action Plan, the city's water supply infrastructure was severely neglected. The Indo-Dutch project installed 71 standposts and 180 handpumps to provide groundwater for drinking and other household uses in the Jajmau district, the district where most of the leather tanneries are located. However, they made no improvement to the Jal Sansthan's river waterworks system. Because of this, middle- and upper-class residents tend to rely on groundwater for drinking purposes. Looking at current figures, groundwater provides over 250 million liters per day (mld) to city residents while the river Gaṅgā provides 120 mld. The Jal Sansthan pulls this river water from an intake point at Bhaironghat, an area lying upstream and off the main bed of the river. The remaining 40 mld is supplied by the Lower Ganga Canal, a separate irrigation channel that flows to the city from Aligarh.

The history of this waterworks system has been complicated by changes in the topography of the riverbed. In 1892, colonial administrators built the first water intake point at a place along the bank of the Gaṅgā at a site called Bhaironghāṭ. When they built it, the main stream of the river flowed just alongside the water intake point. Over the next century, however, the river shifted course. It now runs seven kilometers away from the water intake point. To bring river water to this intake point, the Jal Sansthan has to dredge a channel from the main stream of the Gaṅgā to the pumping station. They must do this dredging frequently enough to ward off invasion by silt, particularly during the monsoon. The Jal Sansthan puts the yearly cost of dredging this channel at thirteen million rupees, a sum that must be paid by the state government since the Jal Sansthan is not able to realize this amount through tariff and tax collection.

To point to the severity of the situation, officials of the Jal Sansthan told me about the disaster of 1962 when the channel was choked with silt and all water supply to the intake pump dried up for two days. This caused a power shutdown because the power station, situated downstream from the water intake point, did not receive any water flow to generate its turbines. The state government called in the military to backflow water from the Lower Ganga Canal, located a few miles downstream, and run it through the water intake station. This provided water to the power station and returned electricity to the city.

After the channel was redredged by the military, an agreement was made between the Kanpur Electrical Supply Authority (KESA) and the Jal Sansthan to dredge the channel for their mutual interests. KESA was responsible for dredging two-thirds of the channel, and the Jal Sansthan was to dredge the other third. Since KESA owned more dredgers and had the skilled staff to perform this function, they undertook the task for the Jal Sansthan against payment. But when KESA shut down its powerhouse on the bank of this artificial channel in 1988, it lost interest in dredging its part of the channel and asked the Jal Sansthan to pay the entire cost of dredging work. Since that time, the costs of dredging have not been paid by the Jal Sansthan or the state government, and a large debt has accumulated. To demand payment, KESA often refuses to perform the dredging services. When this happens silt accumulates and blocks the water flow. In conditions of low water flow, a significant amount of river water is substituted by water from a nearby wastewater drain lying upstream from the intake point. Periodically, the chief minister of the state has to allocate money from other funds to pay KESA and persuade them to continue dredging.

On Noncompliance

In the midst of these institutional complexities, foreign donors and Indian nongovernmental organizations have little success in pushing for better conformity of official practice to policy. The courts, on the other hand, have exerted more influence in this field since the late 1980s. This influence emerged from the work of citizens and lawyers who exposed these disjunctures through public interest litigation (see chap. 7). Court justices helped petitioners to identify the individuals responsible for creating and exploiting public works projects and addressed at a very rudimentary level citizens' demands for more "transparency" in institutional functioning. In some cases the courts (or rather particular justices) made select agencies and officials accountable for these failures by applying financial and political pressure. They issued closures, exacted penalties, and tied the hands of industry executives by limiting their use of public sector agencies and resources. But throughout this, activist justices have been profoundly limited in terms of their enforcement power. Their orders and directions have been effective in meting out short-winded punishments but have not achieved long-term changes in the structure and functioning of industrial units and official institutions.

The Supreme Court, like the High Court and District Courts of the country, can only issue directions to specific institutions to perform their duties. Without their own police to enforce compliance, the courts have to rely on central and state regulatory agencies and civic bodies to implement their orders and evaluate that implementation. Therefore, at the level of practice, they are stymied by the very system of administrators they punish. As the Ganga Pollution Cases suggest (chap. 7), the Central and State Pollution Control Boards created under the Water Act were not thoroughly performing their regulatory functions. In an atmosphere of loose monitoring, industries were allowed to slip in their compliance to environmental regulations and court orders. On the other side, this loose monitoring process encouraged a culture of the mystique in which outsiders, NGOs, and environmental activists imagined that all sorts of vagrancies were being carried out behind the closed doors of these plants. Many independent agencies have alleged that industries are not running their effluent treatment plants; they assume that the Pollution Control Boards rarely conduct surprise visits or report nonfunctioning systems.[14] The Ganga Pollution Cases show that the Uttar Pradesh Pollution Control Board followed Supreme Court orders from 1987 through 1991, but indicate that by 1995 the boards slowed in their regulatory vigilance. This allowed tanneries to develop a system of evading compliance that remained virtually disconnected from citizens' grievances about their functioning. However, the situation of complicity and noncompliance is not that simple. More recently, it appears that, when pushed, the Pollution Control Boards will selectively inspect and report the malfunctioning treatment plants of some industries and use these reports to divert attention away from the many other effluent treatment plants lying idle inside industrial units. In Kanpur, for example, smaller tanners may be targeted and cited for failing to run or failing to own primary treatment plants, while the larger tanners are not mentioned because, some allege, payoffs have been made. Smaller tanners make allegations to suggest that the system of compliance is faulty from within. While it would be difficult to prove the truth of any of these allegations, it is worth listening to them to understand the various interpretations citizens provide about the successes generated through institutional disjunctures. These allegations were made in an interview I had with an owner of one small tannery. The discussion began with references to closure orders made in a High Court case that was running during 1997 and 1998 (see chap. 9). In this discussion, this tanner made references to the ways in which the High Court order to close industries without effluent treatment plants was executed and to the ways in which officials of the Uttar Pradesh Pollution Control Board were behaving. The tanner's comments also reveal the particular rhetorical position that industrialists often take to ward off investigation into their own affairs. The tanner is a young man without much experience in the industry, and he spoke for his father and brother who operate their small family plant. This small tanner discussed his views with me and a member of Eco-Friends, an NGO of Kanpur.

TANNER: Not only me but most of the tannery people, small-scale and big tanners, are annoyed with so many things of these orders [produced in the High Court case]. The reason is that the Uttar Pradesh Pollution Control Board are taking advantages from these orders, in the sense that before giving the approval they are getting the bribe of 40,000 to 50,000 rupees, whatever maximum they can extract from the tanners by making them fearful about the situation. The problems the tanners have with these orders or with the NGOs is that they feel that after installing the primary effluent treatment plants (PETP), this will be a continuous business for the Pollution Control Board to charge them money.

ALLEY: Explain how the process works.

TANNER: First the Court makes the closing orders. According to the orders, then the tanneries are being sealed. When they seal, they put a band with a government logo on the motor and the drum. Raw hide is processed in the drum and all the chemicals including chromium are put in the drum to make the hides soft. They are rotated in the drums several times. The motor which rotates the drum is sealed. In our tannery, we have one drum. Large tanners have seven to ten drums, and small tanners have one or two drums . . . [comments on numbers of drums of specific larger tanneries].

ALLEY: When they come to seal it do they take an appointment or just push their way in?

TANNER: No, they don't take any appointment. When the topic was very hot, they came with a force, along with the DSP (Deputy Superintendent of Police) and a big police force. It is just like a terror to the mind of the people so they can't do anything. They go inside and make the seal.

NGO MEMBER: Let me make it clear. The High Court passed an order on 5 May [1998] that those tanneries which don't have Primary Effluent Treatment Plants should be closed forthwith. So that was the state order.

ALLEY: Because the orders had been done so much earlier . . .

NGO MEMBER: Yes, in M. C. Mehta's case in 1987 and again in 1991 [see chap. 7]. The Supreme Court has passed the order and made the PETPs mandatory for all the tanneries. In a way, all these tanneries that did not have PETPs had been flouting Supreme Court's order [for over ten years]. So the High Court passed a strict direction that all the tanneries which did not have PETPs should be closed forthwith. And the order was to be complied with by the District Magistrate, Kanpur through a high-powered committee under the Chairmanship of Chief Secretary Uttar Pradesh Government. So the PSC and a big force of police—the Rapid Action Force—was deployed because more than 135 tanneries were closed and sealed. In the first phase 70 tanneries were closed and there was discrimination. The Uttar Pradesh Pollution Control Board extorted money and let some of the industries go free.

ALLEY: In the first phase after 20 May . . .

NGO MEMBER: Yes. So again in July there were reports in the newspaper that the Uttar Pradesh Pollution Control Board had taken money from various tanneries that did not have PETPs and still they were not sealed and closed. So this fact was brought to the notice of the Court, and the Court passed an order that if it is true that still tanneries are running without PETPs this situation cannot go on. If it is found then strict action against the UP Pollution Control Board will be taken. So in the second phase, some 60 to 65 tanneries were closed with the police force.

ALLEY: OK, but when they extract the money they are obviously not doing it when the whole police force is there, are they?

TANNER: They don't do it at that time. The force is there when they seal it. The formalities are done after that only. After the installation of the PETP, they [the tanners] will go and inform them [Pollution Control Board] that the PETP is functioning. The UP Pollution Control Board should go there and check the PETP to see it functioning and on the spot give the approval letter. But what they are actually doing is that when the tanner comes to them, they [Pollution Control Board officials] say you [the tanner] should come to our [officials'] house. So they [officials of Pollution Control Board] call these tanner people to their home and indirectly mention that they should come with something. It is like fixed. At first tannery people go through lower people who are responsible for approval. They go step-by-step first to the clerks and pay their 100 rupees to talk to them. Then the paper is being passed upward to the officer and they go to him and say my plant is complete. Then they say "OK right now we don't have time. You come to my home and you'll get the approval letter and this amount you have to pay." It is just a deal they fix.

ALLEY: And the tannery owners know also . . .

TANNER: And they know that if they don't pay this amount, in the future it will always be a problem for them. There is a case in Jajmau. There was a tannery owner, he was a lawyer also. When he discussed this with us he said he was not going to pay this amount. He said it was an order from the Court so after installation of the PETP, approval should be given. He didn't pay. But when he went to the Pollution Control Board and said his PETP is successful, come and check they asked for money and he said he is not going to pay any amount. He waited for ten days and approval was not given. Then again he went to the Pollution Control Board and they said, OK you go and open your seal. He opened the seal but when they came to know that he had opened the seal then they launched an FIR [police report] against him for opening it saying that he shouldn't have. So he got terrorized that the case would go higher. Then he was told that the amount which he is going to pay will be distributed among so many authorities to the higher level. So these people got terrorized. Because such higher authorities are being involved here, they can't do anything. No one is going to help them. So ultimately he lost his hope and again he went back to the Pollution Control Board and paid the amount.

ALLEY: OK, so he pays the Pollution Control Board guy. And is it true that it goes to higher authorities?

NGO MEMBER: It is difficult to substantiate the fact. I also met the guy and I advised him to get the UP Pollution Control Board guy nabbed by the CBI [Central Bureau of Investigation]. I gave him the contact. But he couldn't muster the courage to do so.

ALLEY: But then he would have to be in the act of paying him?

NGO MEMBER: Yes, he'd have to inform the CBI beforehand.

ALLEY: So he probably got scared.

NGO MEMBER: No, but not only one tanner was involved. Hundreds of tanneries were closed and sealed. They could have organized. And they have various associations also. I can believe and I cannot believe them whether they paid or not because I tried to discourage them not to pay. If you have installed your PETP, come with me, I'll go to the Pollution Control Board guy and I will ensure that you get your opening order.

TANNER: See when it comes to the point of whether they have paid, I am 100 percent sure about that because my father has paid that amount.

As this case suggests, the UP Pollution Control Board has a central role to play in establishing standards for industrial behavior, translating court orders, and enforcing industries to comply to court orders. They set the norms for industrial performance and limits on emissions and effluents, inspect industrial effluent treatment plants and their performance, and provide suggestions to industrialists on effluent treatment systems they can purchase or construct. They are at the core of a regulatory machine that does not work to regulate.[15] The personal successes derived from this "regulatory" machine also contribute to the persistence of the public distrust toward water and sewer authorities. The UP Pollution Control Board appears to have the system working in its favor. The mandates upon which regulatory and service agencies are rhetorically based can therefore provide little measure for interpreting their actual functioning. The breaks between discourse and practice are part of everyday industrial and civic culture in northern India.

The point, as Brass puts it in his account of the discourse on riots in northern India, is not to prove the truth or untruth of the allegations (because the truth is ever so elusive), but to outline how the cycle of allegations about corruption and admissions about public alienation runs through wastewater management issues and points attention to specific problems. Citizens and in some cases industrialists as well allege that they do not trust government institutions to provide services without bribe money, and civic institutions do not get the public support they need to improve their services legitimately. In cases of industrial inspection and compliance, the situation is more extreme. Regulatory agencies may be able to take bribe money *and* avoid carrying out their responsibilities, exploiting both ends of the official loop. The political culture of civic institutions does not reward efficiency and self-sufficiency because officials in various departments must answer to the political interests of many other officials who are not their immediate bosses. The system of round-robin favors functions at the expense of executing work in the public interest. The everyday experiences that residents have with representatives of local institutions—who demand bribes to turn a blind eye to regulatory lapses, to manipulate water, sewer, and property taxes and charges, and to threaten those who do not pay—serve to create persistent alienation. While this system of paying for favors is not new, it now hampers the ability of local institutions to solve problems created by industrial pollution and to cater to even greater public demands for water and sewerage services.

To be sure, these regulatory problems are part of a complicated development network, one that pushes industries to produce, export, and make profits, and that gives lip service to requirements to clean up the waste they produce in their core industrial units. According to Schjolden, India's tanning industry has seen a growth in production and an increase in exports through the 1990s (2000:15–20). The support given to industrialists by government officials is part of the larger interest in maintaining an export machine that generates profits for both players through turnover, taxes, and, it is alleged, bribes. At the level of the individual transaction between the official and the industrialist, however, the larger issue of the national good is not the driving concern. Their immediate interest lies in circulating money in such a way that benefits accrue to both parties, which leaves as little as possible for cleaning up the premises and resources they use.

CHAPTER 9

Contesting Water Quality and Mapping Wastewater

BUREAUCRATS ARE NOT THE ONLY GROUPS who map wastescapes for purposes of measuring water quality. Members of NGOs are now taking up the task of tracing out these pernicious flows as a way to analyze the consumption and production practices of industries and municipalities. This is a strategy that some environmental groups and legal activists are using to make allegations about how an industry's or city's effluent treatment plant is functioning on the inside. Since actual data on the internal functioning of these units are not available to activists and researchers, wastescapes must tell most of the story.

As the cases of wastewater management in Kanpur and Banaras demonstrate, local institutions and nongovernmental organizations have not been able to alter the basic distribution of power held at central and state levels in the area of wastewater management and pollution regulation. In policy debates as well, NGOs have had to struggle to get their water quality data recognized by government agencies. In this context, mapping has come to play a pivotal role in arguments between government agencies and NGOs over the ecological status of the river and water quality. Disputes over the validity of government reports on point and nonpoint sources of pollution, on water quality data, and on other ecological reports are often premised upon outlines of wastewater flows and descriptions of the noxious ingredients in them. Since NGOs lack faith in the accuracy of scientific reports put out or sponsored by government agencies, there are times when nongovernmental groups conduct surveys of point and nonpoint sources of pollution and tests of water quality in the same location (and on the same wastescapes) to directly contest the government's data.

Contesting Water Quality in Banaras

In Banaras, local efforts to map wastescapes were initiated more than a decade ago, after three Hindu residents established a voluntary organization, Swatchha Ganga Abhiyan (the Clean Ganga Campaign). In 1982, these Hindu residents launched their campaign from a religious trust called the Sankat Mochan Foundation. This trust acts as the umbrella organization for the campaign, and its acronym, SMF, is generally used to describe the organization run by these Banaras residents. The three directors of the campaign are professors of engineering at Banaras-Hindu University who specialize in hydraulic engineering, chemical engineering, and industrial pollution control. The professor of hydraulic engineering acts as the inspirational center for the organization and is the head of the religious trust that oversees the campaign. He is also the *Mahant* (priestly head) of the Sankat Mochan Temple, a temple situated at the site where the saint-poet Tulsī Dās received his vision of Hanuman, the important monkey god in the Hindu epic the *Rāmāyaṇa* (see Lutgendorf 1991). Since Tulsī Dās's version of the *Rāmāyaṇa* remains a popular classic and a narrative frame within which residents relate their devotion to Gaṅgā (Eck 1982a:87), this priest commands considerable respect from the Banaras Hindu population.

This leader, when describing the early stages of the organization's development, recalled the difficulties involved in personally reconciling the contradictions between, as he put it, his "traditional" and "scientific" views. In a taped interview, he conveyed his sense of struggle:

> So when I started questioning this, that date I cannot say. But I started, I mean, I don't want to, I would say, leave Gaṅgā whom I respect and revere and she is like my Mother. And so I don't want to, just, leave her. And my scientific background says that Gaṅgā water is not clean, it is polluted. But I want to take holy dip in the river because stopping this means leaving the river and I will not. If I don't take dip any day, I think that my day doesn't start. I don't feel, kind of, normal. So I don't want to stop that. And I cannot accept that the

> water is clean. So what should I do? You can at least imagine that this causes a very deep internal stir and pain in my heart, that what to do? What should we do? And actually it is this feeling and this internal struggle and pain and conflict that has given, I would say, birth to the Sankat Mochan Foundation and Swatchha Ganga.

At first, the campaign directors aimed to talk to Banaras residents about the problems of river pollution in the idiom of sacred purity. They framed their instruction with a concern for Gaṅgā's eternal power and purity. When initially developing programs to convey problems of sewage drainage and public waste to the public, the directors sought assistance from American colleagues who eventually established the Friends of the Ganges in the United States. From 1982 through 1985, the SMF developed a local membership and set out to organize school programs, public music concerts, and religious debates to encourage residents to modify their behavior and reduce the amount of household and personal waste entering the Gaṅgā. The president of the women's committee of the campaign lectured on street corners and in women's and men's gatherings about the importance of keeping excrement, soap, and household debris away from the riverbank. They distributed water testing kits in a few schools to introduce the scientific notion of water quality. SMF leaders also directed politicians including Prime Minister Indira Gandhi to problems of river pollution. In late 1994, they worked with the chairman of the People's Commission on Environment and Development, Dr. Karan Singh (the former Indian ambassador to the United States), to host a public forum in Banaras and hear residents' opinions on waste problems affecting their sacred river.

Working in this capacity as local vanguard, the SMF maintains respect for religious notions of Gaṅgā's purificatory power, while pleading for measures to reduce the material waste load on the river. Without stating publicly that Gaṅgā is impure in a sacred sense, for they do not believe this and science provides no proof for it, they maintain the distinction between physical cleanness and purity by arguing that despite her purificatory power, she is becoming asvaccha and needs to be made svaccha. In official circles, using scientific rationality, they publish their water quality data to argue for accountability in sewage management and treatment. While discourse on water quality data does not alienate many Banaras residents, it has not achieved the effect of increasing local membership precisely because residents use different concepts in their assessments and make different connections with survival strategies (see chap. 4).

Since the founding of the Ganga Action Plan in 1986, the SMF has assumed a presence in the debate on environmental pollution in Banaras by taking on the role of watchdog over GAP projects in the city. Initially, the SMF raised questions about the efficacy of proposed infrastructure for wastewater management and later developed the apparatus to monitor water quality themselves. The SMF's earliest proposals suggested that a deep water sewer line be laid on the river floor to avoid the electricity requirements of the pumping stations located along the ghats. In 1992, they organized a conference with funds from USAID in New Delhi to bring together fifty engineers, scientists, administrators from state and central governments, wastewater management personnel, and consultants and representatives from the Central Ganga Authority. Also attending were members of the Massachusetts Harbor Authority, faculty from the University of Stockholm, and consultants with the Thames Authority in London (Sankat Mochan Foundation 1992:14). The delegation discussed the feasibility of this sewer line but government officials concluded that such a scheme would prove too costly. At the conclusion, the invited American guests suggested that the SMF change its strategy to focus on water quality monitoring, to provide an independent data base that could be compared critically with data collected by the Indian Central Water Commission. Other resolutions passed during the January 1992 conference proposed the specification of water quality standards, the delineation of long-term plans for public participation, and short-term steps to improve water quality in bathing areas. Shortly thereafter, a lab was set up on Tulsī ghāṭ, adjacent to Mahantji's home, and equipment and supplies were donated by Swedish members of the conference delegation. Since November 1992, the SMF has followed a sampling pattern that covers five to ten meters from the riverbank and takes samples from a cross section of the river at the upstream region.

In the postcolonial period, assessments of water quality have taken the role that legal definitions of water potability had in the colonial period: attempts to assert the state's right to define the nature of a water resource. In the early stages of the Ganga Action Plan, data were collected and analyzed by government-appointed university research teams. In government-sponsored debates on water quality, officials tried to assume exclusive control over the assessment of water quality and exclude other versions of water quality put out by members of nongovernmental and religious organizations. Under the first phase of the plan, several universities in the Gangetic Plain were awarded contracts to establish water monitoring programs in four cities in the river basin—Haridwar, Allahabad, Banaras, and Patna (Murti et al. 1991). Until 1992, data were forwarded to the Ganga Project Directorate (GPD), which then published results selectively. For educated professors of Banaras who can interpret these results in other ways, the problem is that these data are too controversial to serve as convincing proofs in assessments of environmental pollution. The SMF, for instance, has contested the data passed to the Directorate by academic departments and government water monitoring agencies since the beginning of the GAP monitoring program (see Sankat Mochan Foundation 1990, 1992, 1994; Sharan and Sinha 1988). They have also pressed the GPD to expand its own monitoring program to include fecal coliform count, a parameter the government had omitted in its monitoring schemes.

Evidence of these debates can be found in news media records. Some of the first media reports conveyed the SMF's interest in pushing for more comprehensive water quality monitoring and pointing out the limitations of parameters set by India's Central Water Commission. The *Indian Express* (21 January 1992) reported that government testing of pH value, dissolved oxygen, and BOD (biological oxygen demand) omitted attention to the MPN (most probable number) of coliform bacteria. The press conveyed parts of the debate: "The project authorities in the Jal Nigam had earlier claimed that the high bacterial levels were caused by the dumping of flowers from thousands of temples in Varanasi. But water samples have revealed a substantial presence of human and animal excreta, according to Veer Bhadra Mishra [Director of SMF]."[1] The *Indian Express* presented the government's response to SMF's demands to measure fecal coliform count:

> Animals washing themselves in the river, people bathing, clothes being washed in and around the river add to its coliform content. An official of the Ganga Directorate pointed out that by just bathing in the river, Dr. Misra [one of the Directors of SMF] may be adding as

> much as 100 coliforms. If the river is to be cleaned people may have to take a shower before jumping in for spiritual purification. The NGOs instead of creating an awareness of these aspects prefer to criticize the government, it was pointed out. (21 January 1992)

Press releases citing results of water quality analyses followed. The *Hindu* cited results of the SMF's monitoring of sites along the Gaṅgā, results that showed a total coliform count of 24,000 MPN per 100 ml and a fecal coliform count of 12,000 MPN per 100 ml. BOD levels varied from 8.7 mg per liter to 12 mg per liter at various ghāṭs and was as high as 505 mg per liter downstream from Khirkī nālā.[2] The *Indian Express* reported a joint water-monitoring exercise carried out by SMF, Jal Nigam, the Uttar Pradesh Water Pollution Control Board, and the Central Water Commission that revealed BOD levels over 5.5 and fecal coliform counts of 5,000 to 10,000 per 100 ml.

While participating as a watchdog of the government, SMF members justified their involvement in the debate on waste management and water quality monitoring by referring to the ideal of "people's participation." Blaming government agencies for failing to involve "the people" in leadership roles, the Sankat Mochan Foundation organized public events to create local awareness of sources of environmental pollution. Meanwhile the government also used the "people's participation" rhetoric. Statements by the Union Water Resources Minister appeared several times in 1992 expressing commitment to "people's participation." In the *Hindu*, this minister urged authorities to make the Ganga Action Plan a people's movement by "involving masses in cleaning the holy river."[3] In another release, the press alleged that the Jal Nigam had accused "people" of mismanaging the river. The *Indian Express* wrote, "Jal Nigam has pointed out that despite the policing of the river front, hospital and city wastes continue to be dumped near the bank of the river. Upstream as well as downstream of the river, bodies continue to be thrown into the river surreptitiously."

Through different media, the Ministry of Environment and Forests under the Central Government of India and Sankat Mochan Foundation members denounced the use of soap while bathing and washing clothes and the practice of disposing of dead bodies and animals in the river (Sankat Mochan Foundation 1990, 1992). But these activities, the SMF argued, were not the primary sources of the pollution load. The lion's share of the contribution, they clarified, was created by city discharge: some 250 million liters of diluted sewage running into the Gaṅgā daily. During my own fieldwork from mid-1993 through 1998, I recorded sewage running out of Khirkī nālā (an outfall drain beyond Rāj ghāṭ) and beyond this, from the new nālā discharging treated sewage from the Dinapur sewage treatment plant. The Varuṇā River has also taken heavy discharge from settlements extending along its banks and from the bypass that directs sewage from Konia pumping station into the Varuṇā, which occurs when power is not sufficient to pump the wastewater to the Dinapur plant. Many other drains run from the city down to the Gaṅgā underneath or alongside the bathing ghāṭs. Outfall levels over the past two years have declined in Khirkī nālā but have increased via drains such as the Konia bypass leading to the Varuṇā River. Moreover, the nālā emitting treated sewage from the Dinapur treatment plant takes the 20 to 80 million liters of discharge per day that used to flow from Khirkī nālā. This shift in drains has created an appearance of improvement in the Khirkī nālā drain, the drain that six years ago emitted approximately 100 million liters of sewage per day (Alley 1992:127). However, government agencies are not monitoring and not acknowledging the water quality beyond the outlet from the Dinapur treatment plant. The Central Water Commission's monitoring station remains at Khirkī nālā, not at a point farther downstream beyond the Dinapur treatment plant and nālā.

After the SMF made arguments in many meetings about high fecal coliform counts (FCC), the government agreed to include this parameter and began testing for it in 1994. In their June 1993 report, the Ganga Project Directorate announced that pollution levels in the Gaṅgā were declining as a result of infrastructural improvements in sewage management and treatment. They reported a decline in BOD levels for Banaras from 10 mg per liter in 1986 to between 1 and 2 mg per liter in 1992 (Ministry of Environment and Forests 1993). However, members of the SMF charged that these declines were exaggerated. SMF reports published in 1994 showed higher BOD levels just downstream of Daśāśvamedha ghāṭ at a distance of five meters from the riverbank. In May of that year, BOD varied from 1.11 mg per liter to 26.50, depending on the time of day monitoring was done. Higher figures tended to reflect times when the sewage lines leading to the

Gaṅgā under or alongside the ghāṭs were open (Sankat Mochan Foundation 1994). FCC levels were as high as 440,000 colonies per 100 ml in March and 320,000 colonies in April 1994.

Alternatives and Their Obstacles

In June 1993, the Ministry of Environment and Forests announced that under Phase I of the Ganga Action Plan fifteen treatment plants had been commissioned in the Gaṅgā basin with the capacity to treat a total of 300 million liters per day (Ministry of Environment and Forests 1993). After the treatment plant in Banaras was completed in the spring of 1993, the Clean Ganga Campaign began to notice various operational problems. For example, the plant's collection chamber proved too small to hold incoming discharge from the city during the monsoon. Consequently, the sewage backed up through the main trunk line after heavy rains. During previous monsoons, the city's discharge had been diverted directly to the Gaṅgā, through Khirkī nālā (see Alley 1992:126–27, 1994:132–34). In early 1994, SMF members demanded that government officials close the bypass to the Gaṅgā and divert all sewage to the plant, to comply with the stated objectives of the project. Consequently, the sewage backed up, creating what they call a surcharge in the main trunk line. The SMF pointed to this as proof of the inappropriate design of the collection chamber. In August 1994, after many requests from the SMF, the Ganga Project Directorate convened a meeting of technical experts in the field of sewage treatment to discuss problems with the Banaras plant. The SMF was invited to attend, and I followed along as an observer. At the forum, SMF members charged that a thorough study of city waste discharge had never been conducted before the plant was designed. This is why the collection chamber was constructed improperly. They added that the discharge rates put out by the Ganga Project Directorate in 1994 were also dubious. Pointing out that data from their research contradicted official reports, they demanded a role as an outside witness in future discharge measurement and monitoring. Additionally, SMF members presented a proposal for advanced integrated wastewater ponds as an alternative method of sewage treatment. This proposal had the support of a cadre of foreign experts that included researchers at the University of California, Berkeley and the University of Stockholm. More suited to India's hot climate, they proposed, this method would use only a fraction of the energy consumed by the activated sludge process, reduce BOD levels, and eliminate fecal coliform counts. This alternative would not require greater supplies of energy than those provided by the electricity board of the state of Uttar Pradesh.

Most officials in the Ganga Project Directorate and the National River Conservation Directorate get quite agitated when discussing the main director of the SMF. To them, he is an icon of antagonism, an individual who sticks pins in the ministry and makes his presence known through constant letter-writing and visits. They feel he tries to stir things up through criticism, without providing any constructive work to help solve pollution problems. One director of the National River Conservation Directorate remarked that he does not do all this "for Gaṅgā" but for his own power and status. This is the general charge leveled by government agencies at NGOs. Officials accuse NGOs of participating in environmental discourses only to raise their own status and to claim their hand in government achievements. Brass (1997) has also discussed how this rhetoric of profit making is used by government officials and the police against citizens. Brass's account makes it evident that allegations of corruption work from bottom to top and top to bottom in social hierarchies. That is, while citizens accuse the bureaucracy of corrupt practices, the bureaucracy accuses citizens of the same interests in status and money-making. This is a strategy all sides use to deflect attention away from the actual livelihood strategies they practice themselves.

In 1998, these allegations developed into a more serious standoff between Jal Nigam and Ministry of Environment and Forests officials, on the one side, and the Sankat Mochan Foundation and the local municipality, the Varanasi Nagar Nigam, on the other. In 1997 and 1998, the Varanasi Nagar Nigam passed several resolutions claiming their legal right, as the city's representative body, to choose a sewage management design for the city of Banaras. Several members of the Nagar Nigam were working closely with members of the SMF to propose an alternative sewage management plan modeled on deliberations that SMF had undertaken with foreign consultants. SMF and the Nagar Nigam challenged the Jal Nigam's proposals for the second phase of GAP and submitted critiques of the Jal Nigam's plans directly to the Ministry of Environment and Forests in New Delhi. The Jal

Nigam countered that SMF's plans were unrealistic and costly; they also accused the Nagar Nigam of playing into the hands of the SMF. The mayor, who at one point supported the Nagar Nigam's position, found herself caught in the middle of a power play between the state and the local municipality. The state, which favored the Jal Nigam and its proposal, put pressure on the mayor to withdraw her support for her own constituency, the corporators of the Nagar Nigam. The result was that the mayor grew silent on the matter for almost a year, while the SMF battled directly with the Jal Nigam. The press eventually caught on to the debate and began to report the views of the various parties. The *Times of India* reported:

> The GAP, which was started with a noble task to save Ganga from pollution, had become a bone of contention between the Ganga Pollution Prevention Unit of the UP Jal Nigam and a local NGO—Sankat Mochan Foundation (SMF). There was no sign of "ceasefire" between them over the suitability of measures adopted under the Ganga Action Plan (GAP) for cleaning the river. The first phase of the GAP in Varanasi, which was completed in 1993, should have been followed by the works to be executed under the second phase.
>
> The launching of second phase was yet to be finalised. Both the Jal Nigam and the SMF, which had prepared separate projects for the second phase, were trying to win the race by getting a final approval in their own favour. The allegations and counter-allegations were being raised against each other. According to the SMF chief Veer Bhadra Mishra, the river was as dirty as ever despite the investment of Rs 50 crore. On the other hand, the general manager of the Ganga Pollution prevention unit NC Gupta, held the SMF responsible for the delay in the launching of the second phase of the GAP.
>
> Mr Gupta openly alleged that the SMF, though its proposal was rejected by the experts, was mounting pressure upon the government to get its approval. According to him, the city had been deprived of the benefits of the GAP-II due to the controversies created by the SMF. After the completion of the GAP-I, the unit submitted its project report of the second phase to the national river conservation directorate (NRCD) against which the Union govern-

ment had sanctioned Rs 4752 lakh in 1995. He explained and added, "but owing to the baseless criticism of the GAP-I by the SMF the city could not be benefitted".

Mr Gupta rejected the proposal of the SMF and said that it was neither technically fit nor practical in implementation and the city would have had to face disaster during the flood time if it was executed. Interestingly, he admitted the suitability of the SMF's technology and said "no doubt it is an excellent proposal but practically it is very difficult to be executed".

The SMF's report advocated laying an underground sewer along the ghats which would intercept all the sewage drains. Initially, the Jal Nigam had prepared a report of 12 schemes to be executed under the second phase of the GAP costing Rs 4,752 lakh. The amount was also sanctioned by the NRCD. Later after the intervention of the court, a revised pre-feasibility report of Rs 236.28 crore had been prepared for the second phase of the GAP. The new reports include lying [*sic*] of a trunk sewer, two interceptor sewers, construction of pumping station and sewage treatment plant based on improved oxidation pond technology and upgrading of the existing treatment plants and pumping stations.

Mr Gupta tried to convince that the objective of the GAP-I to divert and intercept 122 MLD (million litre per day) of sewage out of 147 MLD had been achieved. In 1998, the generation of sewage in Varanasi had gone up to 213 MLD which would become 271 MLD by 2013 and 329 MLD by 2028, he explained.

The construction of a new trunk sewer had been proposed in the project of the GAP-II because the existing sewer system, which was commissioned in 1917 for the population of only two lakh [two hundred thousand], was not coping with the increasing load of sewage. On the other hand, the SMF chief, Prof Mishra, a professor at the Institute of Technology (IT) of the BHU and Mahanth of Sankat Mochan temple, was very critical of the explanation of the Jal Nigam officials describing the GAP-I as a total failure and completely inappropriate.

He said that the treated affluent [*sic*] was not only polluting the water sources of the neighbouring villages but also affecting the vegetation of the area. The SMF's project, based on the advanced integrated wastewater pond system (AIWPS) had cost about Rs 100

> crore for the capacity of 300 MLD against the 100 MLD in the present operating system of the Ganga pollution prevention unit.
>
> An 11-member committee constituted by the Union Ministry of Environment and Forest, visited the city recently to review the works done under the first phase of the Ganga Action Plan (GAP). They not only inspected the main sewage pumping station at Konia and sewage treatment plan (STP) at Deenapur but also took stock of the flow of sewage into the river from the NAGWA to the Varuna-Ganga confluence. The committee consisting of the NGOs, environmentalists and public health engineers, had been constituted to suggest measures to make the GAP more successful. According to Mr Gupta, the committee observed that a separate committee should be constituted to examine the projects presented at the NRCD, New Delhi before final approval was given. He said that the committee did not make any remark on the works done under the GAP-I and the operation and maintenance of the plant and pumping stations rather it sought three months more time to make any comment.
>
> Among all these allegations and counter-allegations, one thing was crystal clear that Ganga was getting more polluted on each coming day. Only time could tell what was in store for this holy river.[4]

This standoff is predated by many instances in which NGOs and government agencies have related to one another through an antagonistic relationship. The behavior of the Jal Nigam and the behavior of its colluding agencies was, to the SMF, unacceptable and had to be challenged at every juncture. The Jal Nigam, they argued, was executing a number of questionable engineering projects and spending government money in ways that outside observers had difficulty accounting for.

Profiting from Wastescapes

By late 1995, residents of Kanpur felt that the Gaṅgā's condition was serious. Ten years of the Ganga Action Plan had not produced the results they expected. Many technical projects to decrease the pollution load on the river had been completed or were near completion. But from the point of view of many residents, they had not significantly improved the wastewater problems in the city and in the Gaṅgā. A few active members of Eco-Friends decided that some policing action against the authorities responsible for pollution prevention and industrial regulation was warranted and began to map the wastescapes running through the city and into the Gaṅgā.

In 1995, Eco-Friends conducted a survey of the level of pollution that the visible eye could detect along the riverbank. From these observations, they created a map of the existing drains discharging domestic sewage and industrial wastewater into the river to argue that, contrary to the evaluation report for the first phase of the Ganga Action Plan, many wastewater drains mentioned in the report had not been completely diverted to the new UASB treatment plant.[5] To record the status of infrastructural works projects created under the first phase of the Ganga Action Plan, they visited each of the four pumping stations in the Jajmau district of the city. These four pumping stations are supposed to collect domestic waste and the effluent of over 193 small, medium, and large leather tanneries situated in the Jajmau district (see fig. 10). Kanpur tanneries tan hide with chromium and vegetable dye and discharge wastewater containing chromium sulphide and common salt. Once the sewage reaches these stations it is collected in a chamber and pumped to the sewage treatment plant. When there is no electricity to pump the sewage, it is released from the station into a bypass drain that leads directly to the Gaṅgā.

During their survey tour, Eco-Friends found that three of the four pumping stations had one or two dysfunctional pumps. Without full pump capacity and a continuous power supply, the stations had to turn away sewage and effluent and divert it through bypass drains into the Gaṅgā. In addition to the faulty pumping system, these activists alleged that tanneries were also failing to treat their own waste before dumping it into the city drains.

The pumping station that receives the bulk of the northern city's domestic sewage is located near the secondary treatment plant. With a capacity to hold 160 million liters per day, it receives an average of 120 to 140 mld of domestic sewage. Twenty-five mld of domestic sewage is sent to the 36-mld treatment plant where it is mixed with 9 mld of tannery effluent that is pumped through the four other stations. The remaining sewage load is supposed to be diverted to a larger treatment plant under construction with World Bank funding. For seven years,

Fig. 10. Wastewater flowing from a tannery into a nālā

this sewage load, a load containing high levels of chromium and salt, has been diverted to agricultural fields through an irrigation channel. Some of the untreated wastewater has also been redirected into the flows of treated effluent emitted from the 36-mld plant and the adjacent 5-mld plant. This mix is also bypassed through a channel that runs into the Gaṅgā. During the dry season, the channel gets blocked by sand formations at the riverbank, forcing the effluent into a series of stagnant pools.

Uncremated human and animal bodies were included in the category of gandagī articulated by residents of Daśāśvamedha (see chap. 4). Eco-Friends activists have also addressed the issue of uncremated corpses in the Gaṅgā but differ from pilgrim priests in their assessment of the impact. During boat trips I conducted with them along the Gaṅgā in late 1995 and early 1996, we spotted approximately thirty-five human and fifteen animal corpses floating in the river on each trip. We found skeleton heads lying about or wedged into the sandy banks along the reach of the river at Kanpur. The collection of bones of human and animal carcasses has become a small business for residents who crush and sell them for fertilizer. In Kanpur, the bridge that links the city with Shuklaganj (a town on the other bank of the Gaṅgā) is a popular spot for casting away these corpses. They are thrown from the bridge and from that point are taken farther downstream by Gaṅgā's flow. When the river level is low, the corpses occasionally wash up on the sand banks in the middle or at the edges of the stream (see fig. 11).

Several years ago, Eco-Friends began accusing police officers of dumping uncremated corpses in the river. The police are responsible for picking up these unclaimed bodies and getting them cremated at the local electric crematorium.[6] A small fee is supposed to be paid out of police monies to cover the cost of electrical cremation. In 1997, Eco-Friends alleged in a letter to the High Court of Allahabad that police officers were pocketing the money allocated for cremation and throwing the corpses off the bridge. The NGO activist wrote:

> Hon'ble Sir,
>
> Almost after a fortnight since we cleaned river Ganges of Kanpur by removing 127 deadbodies (117 human and 10 animal carcasses), more than 100 dead bodies could be counted in the same

Fig. 11. Human corpses washed up on sandy banks

stretch of the river. Hon'ble Supreme Court had defined the duties of Kanpur Municipality in 1988 (Ganga Case II). It is the duty of Kanpur Nagar Nigam to ensure that no dead-bodies are thrown into Ganga, but KNN [Kanpur Nagar Nigam] seems to be in a slumber. Our efforts have failed to awaken the government machinery, be it police department (biggest culprit), KNN or Pollution Control Boards (State and Central). Instead, a strange rivalry is going on between Eco-Friends and police department. We're removing the pollutants and police personnel are all out to thwart our efforts and other concerned departments are looking the other way.

Sir, please you do something.

The justice responded rather quickly to the matter, treating the letter as a writ petition submitted in the public interest. He ordered that several respondents be issued notices. By doing so, the justice expanded the notion of locus standi, known in legal terms as epistolary jurisdiction. Rejecting the assumption of personal stake or injury located in the traditional doctrine of standing, the High Court justice allowed this citizen to seek judicial redress under Article 226 of the Constitution and pursue a legal wrong or a legal injury caused to a class of persons who were not able to approach the court for relief (see Mehta n.d.:7–8). The expansion of the notion of citizen standing provides opportunities for concerned citizens and groups to take cases of environmental degradation to the courts. Apart from allowing citizen standing, this particular justice also identified a wide range of the accused by broadening the reading of the letter. In his memo to the court clerk, the justice wrote:

This letter from Rakesh Jaiswal . . . is treated as a writ petition by way of public interest litigation as it involves not only pollution of river Ganges but also of the entire atmosphere at Kanpur. It is further obvious that this position may not only be true for Kanpur but may be equally true as regards the other places wherever there is a city falling on the banks of river Ganges. Accordingly, let the notice be issued to State of U.P., respondent no. 1, State Pollution Control Board, U.P. Lucknow as respondent no. 2, Central Pollution Control Board, Delhi as respondent no. 3 and the concerned authority of Ganga Action Plan as respondent no. 4. The notices

> shall also be issued to Director General of Police, U.P. Lucknow treating him to be respondent no. 5 as the letter under reference states that even when eco-friendly people or society try to clean Ganga by taking out the dead bodies and carcasses the police, instead of appreciating them, thwart their efforts. At this stage, notice is not being issued to the municipal authority as it may not be a matter concerning only one Municipal Corporation but may involve Municipal bodies throughout the state or even outside this state.

Articles 32 and 226 of the Constitution allow a citizen to stand for the public interest and file a case in the nature of mandamus. These provisions are openings within the law for resistance to what is normally considered a part of official hegemony. They are what have made the Ganga Pollution Cases and many other landmark environmental cases possible and successful. But as a judge in the High Court of Karnataka writes, the petitioner must ultimately woo a justice's activism for any concerted public interest litigation to proceed (see Saldanha 1998:13). The justice reading the Eco-Friend's petition was aware of the precedents set in the Ganga Pollution Cases and wrote his court memo to parallel the strategy taken in that Supreme Court case.[7] By broadening the reading of the original writ, he drew in a range of stakeholders whose special interests resided not only in Kanpur but in other riparian cities in the river basin. Following this, the justice appointed a local Muslim advocate to act as amicus curiae for Eco-Friends, but the amicus curiae turned out to be a High Court advocate with no prior experience in environmental cases or public interest litigation.

After the case opened in September 1997 the director of Eco-Friends began a long period of commuting back and forth from Kanpur to Allahabad to argue the matter (previously discussed in chap. 8). What transpired over the next year, until the date that the key justice retired, demonstrates two things. First, the trial record demonstrates that the justice's particular position of advocacy played a crucial role in the success of the petitioner's appeals. But an analysis of events and orders also shows that the petitioner's appeals were subjected to manipulation at another level, outside the legal process.

For citizens and professionals concerned with public interest litigation, manipulations made from the outside present serious challenges.

A report on environmental cases in South Asia poses these concerns in the following ways: How can a court issuing an order such as a mandamus ensure that its order continues to be complied with? What committees of experts can be called in to aid the court, how can they be set up, and what is the reporting procedure they should adopt (Kaniaru et al. 1997:viii)? The following discussion sets the context for understanding why these crucial questions arise.

Before the first hearing in the case, the director of Eco-Friends met several times with his amicus curiae to plan the argument. For Eco-Friends, it was immediately evident that the advocate had no real interest in the matter. Working in a public interest capacity, he had no chance of collecting fees from the petitioner to pay for his expenses in time and office supplies. This meant that the case took a back seat to his other work. In the first court hearing, the amicus curiae was responsible for presenting an outline of the problem, and he did so after much coaching from the petitioner before and during the hearing. Eventually the lawyer was wooed by the newsworthiness of the case, which had become evident to him after the first hearing. This motivated him to continue (though in an uncommitted manner) throughout the next year and gave some strength to the odd client-advocate relationship.

I was present at the first hearing of the case in 1997 and witnessed the initial stages of the development of the petitioner's legal consciousness. By the time I returned to hearings in the case a year later, I found a self-directed petitioner and a sidelined advocate. During the course of the year, the petitioner had become well-versed in the procedures and practices of law in the High Court and had taken many opportunities to argue the matters himself. In that first year, however, the petitioner was also drawn into the operations of his respondents, operations that eventually coopted his amicus curiae. Looking back at the order from the first hearing, we find that the loose working relationship evolving between the petitioner, the amicus curiae, and the respondents was part of the justice's original plan. The justice had instructed in his 15 September order:

> The amicus curiae should communicate with authorities to address grievances of petitioner and masses in general. Litigation is not an adversarial one; full cooperation is expected by all departments to formulate measures to deal with this problem. . . . The sub-

stance of this order is that the issue has to be taken as a common problem.

For the Court, the problem was to decide what outside committees could be called upon to provide scientific opinion and assist in an investigation of government offices and industrial units. Who would act as a "neutral and judicious" player? The justice's aim was to, as he put it, encourage "full cooperation by all departments to formulate measures to deal with this problem." This meant including in the resolution process the government agencies he was sanctioning. This was another very tricky intention! In one particular set of orders and events that evolved from this case, we find that the Jal Nigam benefited from an order the High Court justice passed, much like the Uttar Pradesh Pollution Control Board profited from the case (see chap. 8). A monitoring committee had been set up by the Court to check the work of government agencies. The committee submitted a recommendation that work be done to fix a segment of the sewage system in Banaras, a segment that had been faulty for years. The justice issued an order that money be cleared by the city administrator of Banaras to fix the problem. The Court orders preceding and following this event help to illuminate the past.

The Tapping of Śivālā Nālā

Śivālā nālā is a wastewater drain that runs from Śivālā, a neighborhood lying along the riverbank in Banaras, toward a ghāṭ (also known as Śivālā) and into the Gaṅgā. This drain has been a favorite of mine for many years; every time I go to Banaras, I investigate the status of the wastescapes produced by this tunnel that projects out onto Śivālā ghāṭ from the adjoining neighborhood. Śivālā nālā is situated downstream from the water intake point for the city and upstream from Hariścandra cremation ground and Daśāśvamedha ghāṭ (see fig. 12). The wastewater running through this nālā is a direct indication of the problems generated in and by the wastewater disposal system created under the Ganga Action Plan.[8]

Śivālā nālā was supposed to be tapped under the first phase of the Ganga Action Plan. In fact, all major drains that follow the lines cut by Moghul rulers centuries ago were to be obstructed or "tapped" and the wastewater flow diverted to the main trunk line through pumps and

Fig. 12. Pilgrims going to bathe near Śivālā drain

pumping stations. The main drains in Banaras are Assī, Śivālā, Daśāśvamedha, Trilocan, and Varuṇā, although there are many others that fall between and beyond these points. On the outlets of these old drains, the Jal Nigam installed five pumping stations in the 1970s to trap the wastewater and pump it back to the main trunk line (see chap. 8). An important indicator of the performance of the sewage treatment plant at Dinapur, Śivālā emits wastewater when that downstream plant cannot handle the sewage load of the city, which happens when sewage backs up in the line. When it backs up and begins to seep back into city streets and houses, Jal Nigam officials open the trap doors on these major drains and allow the wastewater to bypass the filled trunk line and flow directly into the Gaṅgā. When Śivālā nālā is most active during the morning hours, it is diverting wastewater that should go through the main trunk line to the Dinapur plant but cannot.

At the 5 March hearing of the High Court case *Rakesh K. Jaiswal v. the State of UP and Others*, the justice ordered that an eight-member citizens' monitoring team be set up to travel to all the riparian cities in the Gaṅgā basin and report on the functioning of government and industrial sewage treatment plants. This committee made its first trips to Kanpur, Allahabad, and Varanasi (Banaras) in March. At the 31 March hearing of the case, the justice commented on an affidavit submitted by the UP Jal Nigam. This affidavit listed work that had not been completed under the first phase of the Ganga Action Plan but needed immediate attention. The justice dictated to the court clerk:

> The Jal Nigam has filed an affidavit of . . . General Manager, Ganga Pollution Control Unit, U.P. Jal Nigam Allahabad-Varanasi. We have gone through the contents of the said affidavit. The said affidavit also mentions the estimated cost in respect of the work which has still to be undertaken and implemented in respect of Ganga Action Plan II. We would like the State government as also the Central Government to let this Court know when this amount which is needed to complete the work contemplated to be done under the Ganga Action Plan I and Ganga Action Plan II is going to be provided and in how many installments this fund is likely to be provided to the Ganga Pollution Control Unit, U.P. Jal Nigam within a period of one month from today so that the work is immediately taken up on priority basis. . . . who is present in Court has

> also been asked to submit a plan to the Court within the said period of one month by which he may be able to satisfy the Court that pending setting up of the treatment plants in what manner the sewer water can be pumped to some destination where there may be large open fields available and where they may be disposed of at different sites turn by turn so that the said disposal at one place ultimately dries up and can then be sold as a manure which may make it possible to save the river being polluted with the sewer being flowed in it and otherwise also making the environment free from pollution. He shall submit his plan in respect of Allahabad and Varanasi as also in respect of such other places where he may think that this system can be made functional. His affidavit shall also indicate the initial amount that may be required forthwith to make this system functional. This affidavit, as indicated, shall be filed within a period one month from today (31/3/98).

In this order, the justice situated himself directly in the middle of the administrative affairs of the Ganga Action Plan and assumed a supervisory role over the allocation of funds and the disbursement to specific projects. On the next court date of 5 May, the justice went further in an administrative oversight capacity. He ordered:

> We have considered the problems of Varanasi and Allahabad particularly in respect of the sewer through which water is being discharged in the riverbank at these two places. We are also informed that Varanasi city at Shivala during the peak season, the treatment plant is over flooded with the result that sewer in Varanasi city at Shivala ultimately comes in the river Ganga. Sri . . . , General Manager, Ganga Pollution Control Unit, U.P. Jal Nigam, Allahabad and Varanasi informed the Court that if an amount of about 2.5 to 3 lacs of rupees (250,000 to 300,000 rupees) is made available to them, then they will see to it that the sewer of Shivala will not overflow. Sri . . . , Mukhya Nagar Adhikari, Nagar Nigam [City Administrator], Varanasi immediately stated that up to 3 (three) lacs of rupees would be provided by the Nagar Nigam, Varanasi to the Ganga Pollution Control Unit if they undertake to do this work. Accordingly, the Ganga Pollution Control Unit, Varanasi is directed to complete this assignment within a period of three months in any case and prefera-

> bly before the onset of monsoon and the report in this respect shall be submitted when this matter is taken up again on 09.7.98. We were also taken through the various problems which are being faced in the discharge of sewer water at Varanasi and Allahabad city. The measures, which are required to be taken forthwith, have been suggested in the court in presence of all the officers. However, it is felt that this matter needs further study to find out whether there can be any alternative arrangement to prevent flowing of sewage in the river within a short period in an effective way. Accordingly, for this purpose, let this matter be listed again on 20.5.1998 by which date not only the Ganga Pollution Control Unit but also the Mukhya Nagar Adhikari of Allahabad and Varanasi city shall let the Court know what are the proposals which can be taken up and implemented immediately by the concerned Nagar Nigam. The Court would also like to know the sources for supply of the materials and their Brand names of reputed manufacturers so that if it may be felt that without wasting for clerical procedure of tender, etc., the required materials be procured the Court may permit the same. (Court Order 5/5/98)

Shortly thereafter, three lakhs or three hundred thousand rupees (ten thousand dollars) were sanctioned for the work, but no other indication about the nature of its completion was provided to the public. Nor was the matter taken up by the justice again in July. When I arrived in Banaras in the middle of September, I began my usual practice of checking the nālā. I found that it continued to flow as it had before, despite the allocation of three lakh rupees to have it "tapped."[9] As most engineers know, individual nālās cannot be tapped unless the larger wastewater drainage system is modified. In the case of Śivālā nālā, the flow of effluent from this drain is directly related to the functioning of the treatment plant situated downstream at Dinapur. When the plant overflows, the wastewater backs up and is released through Śivālā and other drains. The UP Jal Nigam engineers knew that the problem could not be solved with 3 lakhs. So why did they take it, and what did they do with it?

By the summer of 1998, the issue of the utilization of Ganga Action Plan funds had assumed a more central role in this legal discourse. Although the justice had mentioned the issue earlier in his order of 5

March, he did not take it up in a direct way until July. Curiously enough, the justice had initially raised this issue after asking the Jal Nigam to provide a cost scheme for the repairs that needed immediate attention. In that March order, the justice had remarked:

> The court also finds that there are serious allegations that the money received for Ganga Action Plan has not been properly utilized. The suggestion is that the money has been grossly misused or may be misappropriated. It appears desirable to get this probed. Accordingly, it is directed that the parties to this proceeding may suggest the name of some senior retired person or persons of eminence from the Audit and Accounts Service or Government of India who may be provided efficient officer to check and audit the account and submit its report to the court in due time. (5/5/98)

In the hearings of September and October, the justice followed up on this idea and ordered a thorough audit of all Ganga Action Plan projects undertaken in phase one. In the 16 September hearing, the name of the auditor was submitted to the Court. At the 10 October hearing, he appeared before the Court with his requests for infrastructure, personnel, and finances to run the program. The justice met most of his requests and ordered that funds to run the audit be provided by the state government.

After the 10 October hearing, the presiding justice retired. He informed concerned parties that a new bench would be constituted to continue the case. However, by early January 1999, the new bench had met only twice and the judicial will to drive the public interest litigation appeared to die out completely. However, in January 2000, the audit team produced a scathing exposé on the mismanagement and misutilization of funds by the officies of the Jal Nigam, the Vanarasi Development Agency, and the Varanasi Nagar Nigam (*Audit Report on Ganga Action Plan in Uttar Pradesh*).

In a 1998 publication, Justice Saldanha of the High Court of Karnataka explained that the success of public interest litigation is directly dependent upon the activism of Supreme Court and High Court judges. He argued that the last ten years have seen what he calls a Citizen Movement in the courts, a movement that received support from the judiciary from 1990 through 1996. However, recent judicial activity shows a downturn and a reversal of some of the successes of this period. Justice Saldanha wrote:

> The Indian Supreme Court under Justice Kuldeep Singh [the justice hearing the Ganga Pollution Cases] and several of the High Courts initially started rigorously enforcing environment preservation laws and followed this up by handing down some exemplary punishments. Then came an era of unprecedented judicial activism with the Indian Courts virtually playing the role of a father figure in directing the Government agencies to undertake steps toward creation of environmental awareness and curbing large scale devastations and directions to investigate and to prosecute any cases of motivated inaction and a series of orders for purposes of reversing air and water pollution came from the Courts with a degree of regularity. A whole movement emerged and the media assisted to a very large extent by emphasizing the importance of compliance and as a direct result, a sizeable number of public interest litigations came up before the Courts. The lead that came from the Indian Supreme Court had a spontaneous reaction in the major High Courts of the country notably Bombay, Madras, Delhi, Kerala, and Karnataka but then came a sudden set back. With the retirement of Justice Kuldeep Singh, the Supreme Court itself almost stopped entertaining all public interest litigation and even reversed its earlier orders and the High Courts followed suit. All of a sudden, public interest litigation was being virtually shot down and a conscious effort was made from within the judiciary to kill this field of litigation. This has been a sharp set back to the Citizen Movement and a sorry reflection on the judiciary. (1998:13)

The judicial activism of the 1980s and 1990s resulted in many substantive and procedural innovations in environmental jurisprudence. New rights such as the right to livelihood, the right to human dignity, the right to a clean environment, and the right to compensation were advanced through interpretation of Constitutional provisions. There were also procedural innovations: expanding locus standi, admitting letter petitions, allowing suo motu interventions, relaxing procedural technicalities in matters of grave importance, and appointing commissions to collect scientific and other data (Mehta n.d.:6–7).

However, this judicial activism survived to a certain extent upon the charisma of these key justices. Now without those key players, public interest litigation, Justice Saldanha suggests, wears a different look.

The Eco-Friends case also shows how judicial activism can be coopted by the government agencies that justices reprimand and punish and by the powers generated through their collusion with industrial players. I have outlined allegations about how this collusion works between tanners and officials of the Pollution Control Board in Kanpur (chap. 8). In many other public interest environmental cases, court orders focusing on the relocation of polluting industries have played into the hands of government officials, industrialists, and elites.[10] In the tapping of Śivālā nālā, both the justice's and the petitioner's intentions were lost in the results of the order, and the actual behavior of the Jal Nigam remained hidden behind the unchanging appearance of the wastescape.

CHAPTER 10

Transcendence and Immanence

ACTIVE CENTERS OF RESISTANCE against official environmentalism and industrial development, while leading to innovative and substantive legal change, have attracted only a small percentage of the educated elite. Moreover, these resistance discourses work by attempting to enforce the *written* or dominant forms of legal and scientific traditions. This chapter returns to devotees of the Hindu religious tradition and looks in greater detail at their thinking about and relations with government, bureaucracy, and secular policy. This final foray explores Hindu leaders' particular consciousness on waste issues and looks at how they reconcile this consciousness with spiritual commitments. This leads us to the role of the transcendent in the contemporary politics of religion.

Recent studies of Hinduism and ecology emerging out of a crossroads between religious studies, anthropology, and history have enhanced our understanding of the ecological implications of Hindu theological notions.[1] Generally, scholars have considered the notion of divine transcendence in terms of its value in religious or theological discourses and have been less interested in linking these ideas to political processes or material struggles. Sherma (1998:101) provides an eloquent summary of what transcendence has generally meant in religious constructions. She writes:

> Philosophies of transcendence valorize a supreme being, a higher realm, or a state of realization that transcends embodiment, the emotions and connectedness to material life, and discourages a passionate engagement with the world, seeing full participation therein

as a hindrance to spiritual growth. It follows that if the phenomenal world and embodiment are perceived as obstacles, no deep commitment can be expected to this earth or its living forms. At best, a disengaged attitude of "non-harming" *(ahiṃsā)* can be hoped for. A religious tradition emphasizing transcendence would also be expected to give a high valence to spiritual individualism, discouraging a sense of community and the communal responsibility that this partnership with others would entail.

Along with ideas about transcendent power, many accounts of Hinduism have also demonstrated how the earth (Bhu Devī, Bhārat Mātā), rivers (Mā Gaṅgā, Yamunā Devī), and other natural landscapes are feminized in the Hindu tradition, reflecting an alternate value placed on immanent power. Yet as Sherma (1998:96) points out, conceptualizing immanent power in feminine terms may undermine the potential for ecological benefit through the process of sacralization. Generally, the classical Hindu tradition has linked the symbolic feminization of nature with the symbolic materialization of femaleness. It has juxtaposed the nature/material/feminine axis against the transcendent/spiritual/masculine axis and relegated the former to a supporting role. When natural sites and phenomena are feminized, they are also frequently maternalized and become symbolically associated with the complex of expectations surrounding human motherhood. The honor for Mother embedded in this feminization is often based on the Mother's self-negation, her ability to endure privation for the family, to nurture and give sustenance, no matter what the sacrifice, with no thought of her own needs. Sherma (1998:97) explains:

> When the natural world is feminized, the above conception of maternal nature is projected onto the earth or any natural phenomenon that is considered sacred. Consequently, whether it is *Bhūmī Devī* [Mother Earth] or *Bhārat Mātā* [Mother India] or a sacred grove, the expectation is that the sacred site will bless, nurture, purify or perform any other supportive maternal act without any requirement for sustenance in return. Due to such expectations, the potential for ecological benefit to the sacred place is not fulfilled.

When looking at specific traditions in Hindu classical theology, however, the problem becomes more complex. Sherma traces two lines in classical Hindu thinking about women and nature. The first, linked with Sāṃkhya, Yoga, and Advaita Vedanta schools, engenders a dualistic conception of reality: *puruṣa* as spirit and *prakṛti* as matter. Puruṣa is identified with the male gender and prakṛti with the material, feminine principle. Prakṛti is meant to cater to the purpose of emancipating the puruṣa. Nature is distanced as materiality, woman is linked to the impurities of nature and immateriality through her biological functions, and all are rooted in *avidyā* (spiritual ignorance). The ancient notion of *māyā* also supports this association between nature, materiality, and motherhood. Variously described as the power of creating appearances or illusions and the actual material forms of those appearances, māyā becomes the stuff of prakṛti (see Sherma 1998:103; Goudriaan 1978:2–3). In the Advaita Vedanta tradition in particular, māyā condemns individuals to saṃsāra and to the ignorance that leads to suffering. Devalued as *avidyā*, the lower reality of māyā is juxtaposed against *mokṣa*, the ideal of renunciation where knowledge of the absolute Brahman (the divine source of the universe) is secured. Prakṛti must be distanced from the liberation of the self and separated from this true knowledge.

The other line in classical Hindu thinking extends from the tantric tradition, a tradition that is practiced more by lower-caste groups, women, and the marginalized. Tantra makes a similar sort of distinction between transcendent spirit and immanent power but without devaluing nature, materiality, and woman. Especially in tantric Śaktism, Sherma (1998:107) explains, the feminine principle is elevated to reverence for the Great Goddess as the genetrix of the universe. The *Devī Mahātmyā* and other texts in the tantric tradition celebrate the Goddess as the causal agent of creation and the ground of Being. She is the ultimate reality: transcendent as well as immanent, approachable and all-pervasive. She brings together transcendent and immanent power through the notion of *śakti*. As Sherma (1998: 107) puts it, the Goddess is the matrix of the universe and, as *mūla prakṛti*, its material substance. She is transcendent and fully immanent in this world.

It may be rather difficult, however, to identify these two lines of thinking in contemporary popular Hinduism. In discussions about Gaṅgā, both lines do appear present, yet in a manner that suggests complicated revisions in the ways people talk about Gaṅgā, her transcendent power, and her role as a mother (which may be considered her immanent power). Hindu discourses about Gaṅgā embody both tendencies: to exclude all finite relations from the Absolute, making the Absolute totally transcendent, and to include all finite relations by synthesizing them into the Absolute which exceeds them (Sherma 1998:115). They exclude finite relations from the Absolute when arguing that purity can exclude impurity even in the midst of uncleanness. The notion of purity expressed by paṇḍās and other residents of Daśāśvamedha is a transcendent one, not affected by the material and unclean practices of humans and their technologies. These references appear to occur in ritual contexts where men are officiating over communication with the divine (see chap. 4). Yet these informants also include all finite relations when they describe an absorption of impurities and unclean things into the infinite enveloping power of the Great Goddess. These references appear in stories about Gaṅgā as Mother, the provider who can never be paid back for all that she gives. What I would suggest here is that these two lines of thinking occur, not in a theological vacuum, but in direct contradistinction to political power, livelihood interests, and resistances against development and official environmentalism and must therefore be recontextualized in a more dynamic play of significances.

Exploring the extent to which these two lines of classical thinking are revisioned in the contemporary context, I look at the ways several religious leaders express Gaṅgā's transcendent and immanent power. These religious leaders all profess various versions of Hindu faith, but practice that faith and link their discourses about Gaṅgā to their own lives in different ways. One appears more theologically rooted while the other more politically inspired; all suggest a range of expression in the way leaders talk about the transcendent and immanent power of Mother Gaṅgā.

Transcendence and Separate Domains

When nationalist leaders raised opposition to canal works that sought to obstruct and redirect the flow of the river Gaṅgā in Haridwar (chap. 5), their demands to retain the unobstructed flow of the river were not part of an environmental movement but part of a strategy to oppose political rule. These demands invoked religious symbolism to bring religious leaders and nationalists together on a common platform. In the contemporary context, most religious leaders and ritual practitioners do not exhibit a sustained interest in environmental discourse or in discourse that promotes the protection of the river Gaṅgā's purity or flow. They are generally disinterested in activities that promote environmental cleanup or city cleanup projects. Like religious leaders, political leaders have not made great efforts to ideologically exploit the message of *ecological* or *environmental* purity or cleanness, with the exception of the late Prime Minister Rajiv Gandhi. Although religious and political leaders have taken up the issue of "cleaning" or "saving" the river on several occasions and have made references to the spiritual significance of Gaṅgā when explaining the importance of their cleanup programs, they have taken little interest in effectively exploiting this issue in popular politics. Instead, environmental programs initiated by the state are rather lifeless in design and do not connect with the soul of Hindu culture.

Since the inauguration of the Ganga Action Plan in the 1980s, citizens of Banaras, Kanpur, and other cities bordering this river have listened to government statements about pollution prevention and have interacted with officials of agencies working under the Ganga Action Plan. But since the early 1990s, many citizens have become vocal in their criticism of these government programs and skeptical about the successes they tout. Theirs is a skepticism that is difficult to measure, appearing in diffuse ways through local and national media reports, everyday conversation, and commentary from respected citizens and organizations. When this criticism and skepticism is expressed, critics tend to eschew official environmentalism and non-governmental activism as elite, ineffectual practices of hypocrisy.

Despite the general awareness about Ganga Action Plan programs and secular definitions of pollution, by 1995 environmentalists in north India were complaining to me that a curious kind of apathy plagued the public. Others were suggesting that the modern individualism of Hindu culture has diminished respect for public cleanliness, or, as Agarwal (2000:173) has put it, respect for the person outside. Several activists have argued that this "apathy" for the problems of material pollution is most noticeable among religious leaders and pilgrim priests working and living on the banks of the Gaṅgā (on Daśāśvamedha ghāṭ, see chap. 4). To explore the charge further I move to the Māgha Melā, an auspicious festival that occurs every year in the month of Māgha. The melā is situated on a stretch of land adjacent to the sacred confluence *(saṅgam)* of the Gaṅgā, Yamunā, and invisible Sarasvatī Rivers in the city of Prayāg or Allahabad. This city lies upstream from Banaras and downstream from the industrial city of Kanpur. In the month of Māgha (which fell in January that year), I traveled to this saṅgam and interviewed religious leaders and pilgrims who were gathering for religious observance and spiritual meditation. At the saṅgam, respected leaders (Śaṅkarācāryas, svāmīs from large missions and ashrams, and brahmacārīs) as well as *sādhus* (world renouncers) and pilgrims were congregating in a massive tent camp to live by Gaṅgā, cook with Gaṅgājala, do daily ablutions in Gaṅgā, and perform *yajñas*. For an entire lunar month, devotees seek the grace of Gaṅgā and the power to bring about purification of mind and soul.

My discussions with religious leaders at the Māgha Melā revealed that their discourse was distinctly different from the discourses of secular-minded environmentalists and officials of the Ganga Action Plan. Their ways of approaching interaction with the Gaṅgā were decidedly firm, rooted in theology and practice, and prefaced with an understanding of human conduct in this world. Their references to human conduct contained a fair amount of bitterness about politics and official institutions. This bitterness seemed to influence their interest in maintaining a distinction between their religious traditions and the realm of corruption and greed. This interest in distancing their approach to Gaṅgā from worldly pursuits is connected to their lack of interest in protecting, saving, or cleaning Gaṅgā.

I ran the same questions by these religious leaders that I used with paṇḍās and other ritual practitioners at Daśāśvamedha. These questions explored their views on the confluence of two substances: sacred Gaṅgā and liquid and solid gandagī. I was trying to get at whether they thought material pollution affected the sacred purity of the river. When responding, several respected leaders refrained from mixing what they saw as two unrelated subjects: Gaṅgā's transcendent purity and the disintegration of the tangible or material world (represented here by "environmental pollution"). These religious leaders also claimed distance from government agencies and the political process at large by refusing to link the protection of Gaṅgā's cleanness to their own organizational or occupational activities. While many leaders acknowledged that industrial pollution and domestic waste were problems that civic authorities needed to deal with, they did not consider it their responsibility to assist officials in cleaning or protecting Gaṅgā from material pollution. Their role was to oversee the spiritual worship of Mother Gaṅgā.

On the other hand, many alluded to the need for moral regeneration in contemporary society, a regeneration that could have positive effects for civic cleanliness. They stressed the importance of becoming "workers" in the quest for Gaṅgā's grace, casting away desire and practicing nonviolence and celibacy *(brahmacarya)*. They emphasized that these commitments bring one closer to Gaṅgā's grace and to the regeneration of the moral order. A svāmī from Rama Krishna Mission explained:

> We should become workers to obtain the ever flowing grace of Gaṅgā. There should be religious conduct giving birth to truth; birth should be given over to nonviolence, and when one comes for residence *(kalpvās)* on the banks of Gaṅgā one should become a full *brahmacarya*. Stay away from enjoyments or unnecessary things. Do this so your conscience can be pure, so you can be a worker, so you can get Gaṅgā's grace, free from illusion. Like Lord Śaṅkarācārya said, Who is a worker? Who is a worker for grace? They say there are four forms of worship. So what is the first? The first is being full of understanding, the second is having the benefit of sons and service to the cow; the third is the eternal peace *(samādhi)* of saints; and the fourth is salvation. Why? Because we should get salvation *(mokṣa)* from the grace of Mā Gaṅgā.[2]

Although valorizing Gaṅgā's mixing or hydraulic effect, this svāmī and others I spoke with did not bring up the issue of obstructions to her flow, especially those created under government projects to increase hydroelectric power. Instead, these religious authorities argued that Gaṅgā's purity was distinctly removed from transformations taking place in the material or transitory world. One cannot define purity, they seemed to say, by opposing it to the scientific or materialist notion of pollution put out by government agencies and some environmental organizations. The transcendent could not be opposed to, nor (even more important) could it be collapsed into, the temporal or material world. The Śaṅkarācārya from Orissa explained:

> Gaṅgājal gives unparalleled contribution to spiritual and bodily purification. Even scientists give a high position to Gaṅgājal when comparing it with water from other rivers. Water from the faucet or a channel spoils very quickly compared to Gaṅgājal. Gaṅgājal is never spoiled. Gaṅgājal is not affected by the process of degeneration or defilement *(vikritiyo ko dūr kar);* there is a capacity in Gaṅgājal for mixing. So, one should reside at the places where Gaṅgājal has always been pure, and perform religious service *(sevam karnā)* as it should be done according to the śāstras. Gaṅgā must also be pure on the outside. When taking darśan, a devotee, in his heart, cannot see defilement. This is part of the responsibility of having devout sentiment. Also, from the scientific perspective, scientists welcome the importance of Gaṅgājal. So the outer form of Gaṅgā should also be kept clean. They should not be opposed to purity *(śuddhtā).* When one is opposed to cleanness *(svacchtā),* then one's faith in purity falters. This means cleanness *(svacchtā)* is one thing and purity *(śuddhtā)* is another. We make soap from the oil of animals. Some people use it to bathe, to clean clothes. This is the way to make your life clean, but not to make it pure. Where Gaṅgājal is pure *(śuddha),* there, on the outside, one's faith is secured. The eyes are convinced. One's faith is also formed by drinking jala. Therefore it is important that we pay attention to her cleanness. In whatever place *(sthān)* Gaṅgā is important, and wherever Yamunā is important in her place, and wherever the hidden form of Sarasvatī is important, and where all three meet at a confluence *(saṅgam),* this conception is manifested in the Vedas. Therefore, from the spiritual perspective, we worship the boundless importance of the saṅgam.[3]

Indeed, my questions set the respondent up for the issue. I asked the respected Śaṅkarācārya if he recognized a distinction between purity *(śuddhtā)* and cleanness *(svacchtā).* To this, the Śaṅkarācārya responded by arguing that purity *(śuddhtā)* was associated with Gaṅgā's transcendent power and cleanness *(svacchtā)* with her outer form *(bāhar se).* Gaṅgā's transcendent purity could not be opposed to or collapsed into her outward form. Rather, he explained, the two were considered necessary complements of one another. The key to this view is embedded in the phrase *vikritiyo ko dūr kar* (distanced from defilement). By this, he asserted that Gaṅgā is removed from and thus not affected by defiling elements. Taking this point further, the Śaṅkarācārya explained that Gaṅgā had different forms to deal with the separate domains of transcendent purity and degrading material. Her outer or material form was important to a pilgrim's faith because it convinces through visualization. The outer form figures into Gaṅgā's purity, making cleanness *(svacchtā)* complementary but not conditional to purity *(śuddhtā)*. Yet the Śaṅkarācārya also alluded to political and secular domains, suggesting that he defined his sense of the transcendent and the material against contemporary social and political issues. He continued:

> But about the matter of science, our religion is not opposed to it nor uncongenial toward it. We support scientific research. Like the matchbox. . . . But in our active moral work *(kriyā karm)* we will respect our own dharmaśastras. In worldly pursuits to the point that science is helpful, I welcome the usefulness of science. Scientists have created electricity; they have given computers, rockets. We don't oppose this. Opposition occurs when there is conflict *(takrār)*. They are not one subject but separate subjects. The Vedas gave special importance to one subject and scientists did not give importance to this. Our śāstras dealt with the subject of behavior and our śāstras have freed us from bodily suffering and worldly affairs. Because in the end, in plight or prosperity, these are the two benefits of our faith *(dharm)*.

Therefore, separate expertise is required to understand each realm. He continued:

> For example, we can ask, do you believe in your eyes or your ears? Do you understand? To see the essence of form, do you believe in your eyes or your ears? What answer will we give? In our eyes! Because the understanding of form is based on the eyes. The eyes are supreme in relation to form. So someone may say that in regard to all sensation *(sparś)*, some worship with the eyes and others do not. If knowledge of sensation is possible in other ways than through the eyes, then isn't the subject of science separate? The subjects of religion *(dharm)* and the śāstras are separate *(alag)* from science. When both subjects are separated *(alag alag)*, then the contributions of

> both in their own fields can be appreciated. The lawyer is important in the court; the astrologer is important for making horoscopes. Both are important in their own right. But when we say that an astrologer should do the lawyer's work or if we tell a lawyer to become an astrologer it will not work.

Advocating the separation of religion and science rather than the opposition between them, he gave religion a transcendent character because it could not be equated with or challenged by science. Practitioners of each profession have their own rights and duties that must be appreciated and protected; one cannot do the work of the other. Without directly referring to the problem of power, this esteemed leader implied that the integrity of each expertise should be based on respect for each sphere of influence.

This sense of separation was also echoed in the casual complaints of many ritual specialists on Daśāśvamedha ghāṭ when they argued that cleaning was not their work but the work of the municipality and of state agencies such as the Jal Nigam. Yet the paṇḍās' sense of separation is slightly different from that expressed by leaders at the Māgha Melā. The distinction between occupational specializations made by paṇḍās can be taken as a metaphorical extension of the specialization of duties within the caste hierarchy. The principle of nonoverlapping functions that Marglin and Lingat have previously described might be applied to this understanding of separation.[4] In Marglin's discussion, for instance, separation and wholeness are relevant in the ascending order of caste: when inferiors cross caste boundaries their behavior invokes pollution, chaos, or confusion. Thus high-caste groups attempt to keep duties, food, articles, and other personal effects separated from the reach of lower-caste groups. This separation allows them to appear autonomous and independent from other groups, and thus transcendent. It is also possible to take this sense of separation less abstractly, by understanding it as a strategy upper-caste Hindus use to reject contact with lower-caste occupations such as sweeping and cleaning that deal with impure substances (Korom 1998). But in thinking about the Śaṅkarācārya's position, other complications arise. While he may be referring to occupational specializations, he is doing so to support a theological position, which then in a more intellectual way can be used to explain the distinction between purity and the material world. This is a distinction that might be useful for others wishing to reapply it to worldly affairs in other contexts.

This is what another respected leader did a year earlier, when delivering a lecture on the birthday of Mother Gaṅgā. Swami Chidananda, leader of the Divine Life Society, spoke to a group of devotees and environmental activists who had gathered on Gaṅgā's birthday (Gaṅgā Daśaharā) in the Himalayan town of Gangotri, just downstream from the mouth of one of Gaṅgā's tributaries. At that time, Swami Chidananda promoted an activist consciousness and practice toward Gaṅgā. Likening this to the direct action of Mahatma Gandhi, the respected svāmī invoked the soul force Gandhi derived from within to seek direct communication with God. Making Gandhi's commitment a model, he advocated lifework without any connection to government, or what he called the *jyośī-sarkar* connection. He argued that this kind of active lifework was necessary in an environment in which an antireligious state was attempting to undermine Hindu life and culture. Passive encouragement of this trend, he warned, would only result in cultural decline. Only with mental vision and devotion *(bhaktī-bhāv)*, the svāmī explained, can one work for Gaṅgā and reap results *(prabhāv-samarpaṇ)*. Swami Chidananda spoke about Gandhi on that day:

> He [Gandhi] had a soul force, the conviction, willpower and resolve to free the motherland. This is the spirit, the driving force behind his frail physique. We should generate this soul force from ourselves, from within, and that force comes from God. For Gandhi, that connection with God was the Gītā. When he was reading the Gītā he was in direct communication with Lord Kṛṣṇa. It is by connecting ourselves with the cosmic infinite, illimitable, boundless, immeasurable divine power which we call God that we will have the power to persevere and succeed. One needs to work the whole life.

Swami Chidananda pointed out Gaṅgā's role as a lifeline to Indian culture and civilization and encouraged a spiritual dedication to cleaning the river. This would be an activism matching that of Bhagiratha, the descendant of King Sagara who did spiritual meditations *(tapasyā)* for years before Brahma agreed to let Gaṅgā down to earth. But while highlighting direct communication with God and spiritual commitment to seeking truth, Swami Chidananda advocated separation between the people and government by making a more direct reference to power and struggle. He said:

> People say they come to destroy mosques and temples. At that time [of the conflict in Ayodhya], there was so much passion generated. In one day, from the Himalayas to Kanya Kumari, from Arunachal and Meghalaya to the Punjab, they carried this passion. Why don't they have this much passion for saving the Gaṅgā and for the movement to save other rivers for the next generations? Why not? I have said that passion should be applied because humanity takes direction from love or hatred. This is the petrol and diesel used to move men. There in Ayodhya the passion is directed against Muslims. They are engrossed in anti-Muslim sentiment. It is a political thing. So everyone works for this passion. But here there is not this kind of passion because if opposition [*takrār*] occurs, it must occur with the government. But at every step we need the grace of government: for a permit we need the government, for information we need them, for everything. We don't want to need the government! It's like that. But there is a concept that we don't need any pomp and show from government. Go ahead and take what you want from government but I am ready for a hunger strike. *I don't want anything from sarkār [government]! Then why should I care for sarkār?* When we want something from them then our internal structure becomes weak. When we don't need anything we are strong. So without desire, with connection to God, we can sacrifice our own lives.[5]

Swami Chidananda alluded to the pervasiveness of Indian bureaucracy and the embeddedness of bureaucratic power in everyday life. While rejecting government assistance, he proposed that activism proceed outside the web of bureaucracy. Since, as he put it, a "blind and non-comprehending attitude in the minds of the bureaucracy" was preventing real communication and undermining Hindu life and culture, *bhaktī-bhāv* and *viśvās* (belief) were needed to create work with moral vision.

While environmental activists find this vision theoretically acceptable, they have also commented to me that another more unfortunate response often arises among the public. These calls for opposition to bureaucracy have had more success in encouraging a kind of passive resistance to government power. This, activists point out, is evident in the positions taken by the paṇḍās of Daśāśvamedha and by other ritual specialists in sacred places along the riverbank. Religious specialists with everyday connections to Gaṅgā often blame the government for its failure to clean the river under the Ganga Action Plan and for perpetuating institutions that are unresponsive to citizens' needs. This appropriation of blame is a kind of everyday resistance for them. Yet when it comes to advocating active reform of the kind suggested by Swami Chidananda, the subject stops, and blame is directed at government officials who block citizen action. It appears that connecting Gaṅgā's transcendent purity with the necessity for active worldly reform is a complicated task indeed.

A significant stumbling block emerges from the fact that the tradition of religious mobilization has never encouraged engagement in any form of *environmental activism.* Eulogies to Gaṅgā and worship of Gaṅgājala are central sacred symbols in pilgrimage and in pūjā, but they do not describe Gaṅgājala as a finite resource with contours shaped by a larger ecosystem. Moreover, many Hindus would not confuse religious rituals with civic ethics. Pūjā and pilgrimage are not undertaken to generate social effects but are the means to communicating with and seeking grace and power from a divine force. The presence of this divine, immanent force predetermines any social effect that may arise from ritual.

The historical material outlined in chapter five showed that the political appropriation of purity symbolism could be included in a resistance movement against colonial power. This was an appropriation that served the purpose of mobilizing religious and political leaders under a common political cause; it set up a precedent for the political use of purity rhetoric. Yet this early-twentieth-century example did not evolve into a movement of any kind and has not provided a firm basis for legitimating mobilization in the face of contemporary problems. Since the appropriation ended in an agreement between government, scientific, and religious communities, it failed to provide any lasting condition for resistances surfacing today. Gaṅgā's purity has lapsed into a fixed transcendent state.

There are, in other words, many religious leaders and ritual specialists who are explicit transcendentalists and deny any possibility of an upper limit to Gaṅgā's purifying power. For them, river pollution is an altogether erroneous notion, and Gaṅgā's powerful flow and motherliness attest to its fallacy. As the Śaṅkarācārya told me, eternal power *(acyuta viṣay)* cannot be understood through the eyes that affirm science. This is because knowledge comes from direct communication with God. Knowledge *(jñān)* of Gaṅgā cannot be collapsed into knowledge of material transformations. It cannot be approached in the way one washes dirt off the body with soap. It must be done through a holistic commitment to dharma. Scientific treatments to "clean" the river are only as good as soap; they cannot reach or transform divine power. Cleaning the river and gaining knowledge from Gaṅgā are separate pursuits and should remain in their place.

Many religious leaders also explained how the vectors of space and time *(deś aur kāl)* intersect with the manifestation of sacredness. This is the visual truth *(darśnik tathya)* that underlies Gaṅgā's importance. For example, Gaṅgā's power is especially apparent at Prayāg during the month of Māgha, on the last day of the dark half of the month *(amāvasyā)* and on the full moon day *(pūrṇimā)*. Her power intersects with radiant place *(ujjval deś)* and radiant time *(ujjval kāl)*. In this view, Gaṅgā's power is not intermittent and has no upper limit or capacity. If rituals make her power appear intermittent, it is only because that power must be accessed in radiant place and time. Gaṅgā occupies a fixed sacrality beyond space and time *(acyut)*. Therefore, she is imperishable in both space and time.

The Immanence of Mother Gaṅgā

It was at Swami Chidananda's lecture on the auspicious occasion of Gaṅgā Daśaharā that I was fortunate to meet the head of the NGO Eco-Friends for the first time. A year later, in 1995, this Kanpur-based activist decided to approach other religious leaders and encourage them to take up the call to save or protect Gaṅgā from material pollution. He did this by attending religious celebrations and by approaching leaders of various organizations. On one occasion, I accompanied him to the Delhi office of the Vishwa Hindu Parisad (VHP), the World Hindu Organization, where he met with several of the organization's highest administrators. Approaching this national organization, the Eco-Friends activist hoped that they would take his suggestions seriously and adopt a call to save the sacred river Gaṅgā from material pollution. During that meeting, the VHP officials agreed to include this message in a pilgrimage rally scheduled for later in the year. They added Mother Gaṅgā to their list of sacred symbols and hailed her powers of purification in organization memoranda before the start of the rally. However, during the actual processions the message was only lightly publicized. In effect a nonstarter, the issue failed to attract the attention of leaders of the VHP or members of other religious organizations. Apparently they saw little political valence latent in or already created by the government's call to "save" or "protect" Gaṅgā from pollution.

This pilgrimage rally, called the Ekatmata Yatra (All India Harmony Expedition), originated in several different parts of the country and gathered supporters from across the subcontinent. Branches of the pilgrimage started at Rameshwaram and Kannyakumari in Tamil Nadu, Pashupatinath in Nepal, Gangasagar in West Bengal, Somnath in Gujarat, Parashuram Kund in Arunachal Pradesh, Amritsar in Punjab, Mahavirji in Rajasthan, and Haridwar in Uttar Pradesh. They cut paths across the landscape, some of which followed the riverbed of the Gaṅgā for short distances, before converging in Nagpur, Madhya Pradesh. In Nagpur, many VHP leaders gathered for a final rally.

The VHP's stated aim was to teach Indians about the importance of Mother India, Mother Cow, and Mother Gaṅgā and, as they claimed, bring unity to a Hindu community that had "fallen asleep." Their printed material stressed the importance of the cow for agriculture and Gaṅgā for the advent of Hindu civilization. But these themes were really meant to set the context for resurrecting their 1984 promise to destroy Muslim mosques in Ayodhya and Kashi and a Muslim Idgah (place for prayers on the Id festival) in Mathura. To these sites of Muslim worship they linked rivers held sacred by Hindus: the Sarāyū with Ayodhya, the Yamunā with Mathura, and the Gaṅgā with Kashi (Varanasi). Mother Gaṅgā, the archetype of all rivers in India, acted as the key symbol of these associations. The mandate they published for the event stated:

> The presiding deities of all these *yātras* will be the three Mothers, Gaṅgā Mātā, Bhāratmātā and Gowmātā, whose images will be adorning the motorised chariots. In these *yātras*, there will be

> chariot programmes, worship of Gaṅgā *kalaś* (jug), worship of cow-alongwith-her-calf [*sic*], and worship of the sacred soil of India (sacred soils of Sri Ram Janma Bhumi, Ayodhya, Sri Krishna Janmasthan, Mathura, and Sri Kashi Vishwanath, Gyanvaapi, Varanasi). These worships will be conducted in 300,000 villages. There is the possibility of this number increasing.[6]

To initiate one branch of the yatra, VHP leaders gathered with svāmīs, Śaṅkarācāryas, and members of Parliament on Har-kī-paurī ghāṭ in the sacred city of Haridwar.[7] There they performed several pūjās. At their conclusion, pilgrims and tourists took darśan of the spiritual leaders and registered their moral support. But as the yātra moved away from the riverbank and across northern India, the public discussions turned to the goal of altering the landscape of Islam and barely mentioned the call to save Gaṅgā from material pollution. This was despite the fact that the objectives of the yatra explicitly included the aim to "build up assertive public awareness to ensure maintenance of sanctity and purity of the Gaṅgā; and also awareness against environmental pollution." In all actuality, this concern took a back seat to the interest in reinforcing the idea of the Hindu community's Muslim enemy and using this identification with a common hatred to fuel the yātra. The latter interest never appropriated the former, leaving the concern for Gaṅgā a nonissue.

The interest in demonizing Muslims and issuing calls to reappropriate Hindu sacred spaces controlled by them was not linked to the environmental concern about the river because an easy transference could not be made. A Muslim enemy could not be created to mobilize audiences behind the call to save the Gaṅgā. While Hindu nationalists cast Muslims as the demonic other in struggles for sacred space, they have no basis for treating Muslims as enemies of the river. Although one could argue that Muslim industrialists pollute the river, Hindu industrialists do the same, and the two groups are often little different in their behavior. Moreover, Muslims do not behave disrespectfully toward the river. They also bathe in the Gaṅgā and offer religious donations *(dān)*, especially in the form of fish *(maclīdān)*. Although I have heard Hindus in Banaras complain that Muslims consider Gaṅgā a mere river (by calling her *nadiyā*), I have never seen Hindu citizens focus on a hatred of Muslims as a way to defend

Gaṅgā's purity. So if the VHP had any interest in an environmental message, they had little chance of using the sentiment wrapped up in their opposition to the Muslim other as a motivating force. And since most of their activities used this motivating force for political objectives, they did little with antipollution rhetoric.

That said, I must turn now to more recent examples in which religious and political leaders have appropriated antipollution and antidam rhetoric. But as the examples will show, these have been rather brief political appropriations that, again, reveal the weakness of this rhetorical power among public audiences. In very recent political campaigns, there has been some mention of "flow rhetoric" by leaders who tend to regularly weave religious concerns into their political doctrine and practices. In the 1998 parliamentary elections in India, VHP and BJP leaders briefly issued calls in party newspapers and memos to stop the building of the controversial Tehri Dam in the Garhwal Himalayas. One paper read:

> The religious importance of the pure river Gaṅgā *(pavitra Gaṅgā nadī)* has remained with citizens along her banks. But with the advance of science come consequences and now it is very unfortunate that the natural flow *(prakritik pravāh)* of Gaṅgājal will be badly affected by the formation of a stagnant lake. In the world, wherever dams have been made to stop the water of a river, environmental problems have developed. For example, there is the Aswan Dam. After the construction of the dam, the water of the river became polluted *(praduṣit)* and new kinds of illnesses were born.[8]

Opposition to the dam and the valorization of unobstructed flow are not new to Indian politics (see chap. 5), but they are issues that many postindependence leaders have been reluctant to raise in their own campaigns. These surfaced briefly in the 1998 elections, I would venture to say, because of the perspective and concerns of a particular sādhu-sant, a religious-political leader who plays a prominent role in state and national politics. Years earlier, this sādhu-sant was approached by an activist from Eco-Friends, who may have had some effect on his thinking. This is an interesting connection because it suggests that an NGO's variant of environmental activism has, on some occasions, made headway in religious and political circles. But as this case will show, it made only limited contributions to political or religious movements and continues to remain peripheral to other campaigns organized by leaders who use religious values to construct political ideologies.

This particular sādhu-sant is Swami Chinmayananda. Leader of an ashram and member of the Bhartiya Janata Party, Swami Chinmayananda has on several occasions incorporated the theme of environmental pollution into his religious sermons and political campaigns. In the early 1990s, he was elected member of Parliament from Uttar Pradesh and, after losing that seat in 1993, began to lead rallies to publicize the problems of pollution for the Vishwa Hindu Parisad (see Van Dyke 1997, 1999). This svāmī's concerns were influenced by discussions with the Eco-Friends activist, though these concerns have not generated the energy to sustain a rhetorical campaign for a long period of time. Swami Chinmayananda's public display surfaced in the summer and fall of 1997, when he was campaigning as a political candidate for the 1998 general elections. In the fall of 1997 Swami Chinmayananda led a rally to promote specific VHP and RSS (Rashtriya Swayamsevak Sangh) issues, among them an interest in giving a push to the Rām-Janmābhūmī movement. He explored the idea of expanding upon antipollution rhetoric by forming the Gaṅgā Rakshā Samiti (Committee for the Security of Gaṅgā), of which he was nominated president. The committee brought up this issue in conjunction with the Cow Protection movement and the Bhārāt Mātā concept. All three issues were to feed into the resurrection of the Rām-Janmābhūmī movement. For several weeks, Chinmayananda and several other VHP leaders and BJP candidates journeyed by motorized launch along the Gaṅgā, starting in Calcutta and landing in Allahabad.[9] The activist from Eco-Friends also accompanied the entourage and supplied me with many of the reports about the pilgrimage.

During the tour, many of the speeches made by participants publicized the dangers of the Tehri Dam and made comments on the condition of the Gaṅgā. Chinmayananda's calls to protect Gaṅgā's purity were well articulated and documented. In a report he wrote on the events of the yātra, he explained that their aim was to make Gaṅgā free from pollution *(Gaṅgā ke praduṣan mukt banāne)* and to save the soil, fish, other animals, and communities along the river's edge from the same fate. In several places in the report, he summed up his observations and reactions.

> Having seen the swelling torrent of people who welcomed the *yātra*, our enthusiasm grew and we were happy. But then seeing the plight of Gaṅgā *(Gaṅgā kī durdśā)* our hearts were anguished. The dirty and smelling drains coming from the riverbank towns, the dangerous waste of industries, the floating corpses of animals and men and among this a crow hovering above looking strangely contented. In each place, boatmen, fishermen, pilgrimage priests and bathers and riverbank farmers were troubled. In this long *yātra*, we saw every dirty drain in every town, every hamlet. Even where schemes had been started to stop the dirty drains, we could not see any improvement from efforts to treat the wastewater *(jal)*. We were shown places with electrical crematoria, but they were all closed down.[10]

The importance of Gaṅgā, he continued, lies at the heart of national unity and the social battle for truth.

> In fact, Gaṅgā has an extraordinary capacity *(adrabhut ksamtā)* for national unity and battles over truth. Forty crore [400 million] people are connected to Gaṅgā. In the courts of justice, an authoritative oath on *Gaṅgājal* is professed. Every year crores of people take jugs of water and do *jalābhiṣek* in Śivā temples around the world. In the world there is only one *jaldhārā* (flowing nectar) whose water has qualities that astound scientists of the world.

The benefits, the svāmī added, are not limited to one group.

> From Gomukh to Gangasagar in its 2250 km journey, she created numerous sacred places which are well appreciated in Indian culture. All the organizations and religious groups welcome the grace *(divyatā)* of Gaṅgā. Followers of every *jāti*, sect or religion are worshipers of Gaṅgā. So this is the situation, this is the Gaṅgā who is connected with the whole population of India, whose purity *(pavitratā)* should be protected by the Indian government and who should be established as a national heritage trust.

In these passages, Swami Chinmayananda clearly stresses Gaṅgā's immanent power and the totalizing role she plays not only in Hinduism but in Indian civilization and national unity. She is the ground for humanity and the means for religious liberation, the immanent and the transcendent. Yet Swami Chinmayananda went on to criticize the government for failing to improve the cleanness of the river, hinting at its misuse of power and corrupt practices. He claimed that the decline in the condition of her cleanness *(uski svacchtā kī dasā)* was evidenced in the rise of BOD levels. Gaṅgājal, he said, is not fit for drinking beyond Haridwar, and Tehri Dam will only make this condition worse.

In 1998, Swami Chinmayananda was elected to office, and the campaign against river pollution and the opposition to the Tehri Dam fell away almost immediately. Neither protection of Gaṅgā's purity nor opposition to the dam resurfaced in his political ideology again until early 2001 when government officials were preparing for the impoundment of the Tehri reservoir. Despite his very recent resurrection of the issue, this concern for Gaṅgā has not been used to manipulate official bureaucracy in any significant way, nor has the short-lived campaign against river pollution become an instrument for mobilizing the public or promoting official—that is, private—projects of national development. While Swami Chinmayananda's references to Gaṅgā's transcendent and immanent power may have reached the public audiences they addressed during the pilgrimage, they were not effective in capturing a lasting rhetorical space in politics or a link to the electorate by creating public expectations for change. In particular, the references had no real valence in encouraging secular forms of interchange with the river or in sustaining the ideological tryst between religious and official understandings of the river's power. Swami Chinmayananda's directives proved to be short-lived for different reasons than the Hindu nationalist movement's described in chapter 5. While expedient for momentary political gain because they connected with the Hindu soul, they had no long-term effect for bureaucracy or official environmentalism where connections with the industrial lobby play a more important role. Once in office, by discontinuing the campaign to protect Gaṅgā, Chinmayananda made the implicit gesture that references to Gaṅgā's immanent power and her role in religious consciousness, civilization, and national unity were no longer necessary to the maintenance of his own power.

Swami Chinmayananda's references were slightly different from those made by the religious leaders at the Māgha Melā. While the latter tended to stress Gaṅgā's transcendent power and the need to distance pursuits for salvation from the worldly activities of cleanup and civic action, Swami Chinmayananda made a call to broach the quest for the transcendent with activities directed at cleaning the river. This made sense for his political interests: by likening Gaṅgā to the soil of the land, to the national spirit, and to the ground of humanity, he could make her integral to the way he envisioned his own constituency and made claims to represent them. By advocating protection for Gaṅgā, he was promising to protect his constituency, his people, and his nation. But Gaṅgā's immanence proved just as unlikely as her transcendence to galvanize a cleaning movement. The distinction between purity and impurity is still persuasive, but political temptations and strategies for capturing and maintaining power pull leaders away from interests in waste management for urban ecological improvement.

General elections from 1989 through 1999 brought sādhu-sants into politics at state and central levels (see Van Dyke 1997, 1999). But when campaigning, these sādhus participated in only select domains of public discourse and interaction. Using their powers to bless others, sādhu-sants drew their own benefits from political power. Likewise, the mystical powers of the sādhu were exploited by politicians who used them as agents of mass mobilization. But as more recent accounts of the decline in their power are making known, their realms of influence have always been selective and limited.

Their selective and limited powers are revealed by the fact that Cabinet or high administrative positions in environmental programs for cleaning the Gaṅgā have not been filled by sādhu-sants. Planned by the Central Ministry of Environment and Forests and executed through state governments in north India, the Ganga Action Plan by and large employed secular-minded bureaucrats and technicians to set up waste disposal and treatment infrastructure in all the large cities bordering the Gaṅgā. Officials involved a large number of engineers and technicians and departed in all ways from rules of waste management embedded in religious traditions. The extreme scientific stance and the emphasis on technological solutions to waste problems made residents and sādhu-sant politicians alike lose interest.

After the 1998 elections, VHP activists and BJP politicians abandoned the save-the-Gaṅgā cause and opposition to the Tehri Dam because their goals for political mobilization did not pan out. A sustained opposition to the Tehri Dam and a sustained concern for pollution of the Gaṅgā would have put their own BJP government in an uncomfortable position, by undermining support for the very constituencies they were supposed to represent. Instead when faced by political instability, VHP activists and BJP politicians developed a new scapegoat to aid in the reintegration of their political power. They turned their focus toward Christian minority groups, and later the Pakistani government and Muslim militants, as they looked inward at the definition of national identity and outward toward the border conflict with Pakistan as a way to chart their political path.

Transcendence without Encompassment, Immanence without Activism

By defining Hinduism and science as separate areas of specialization, religious leaders at the Māgha Melā appeared to elide contradictions between the cultural logics of Hinduism, science, and secular politics in order to minimize ideological conflict at a time when resources are in great demand. Yet we should examine these statements in the context of their assumptions to power and authority. These statements support an ancient institutional complex, one that has organized access to sacred power for centuries. Official institutions are now considered the legal and political guardians of national resources, and statements of religious leaders can be taken as attempts to recapture ancient rights to access in the face of secular claims. At the official level, departments under the government of India and the judiciary deal with resource management. Yet religious leaders and institutions—Śaṅkarācāryas, sādhu-sants, sectarian organizations, temple committees, trusts, and public service agencies—are also significant players, not recognized by government as legitimate resource managers but respected by the public as moral, cultural, and, at times, political authorities. Government agencies regulate uses of the river at the legal level by restricting navigation, fishing, the building of dams or other obstructions, and the extraction of water in specified zones. Empowered by the laws of the

Pollution Control Boards, officials can exact penalties for dumping waste and effluent. However, legal restrictions do not provide the only codes of resource management understood and respected by the citizenry. Procedures for performing ablutions in the sacred texts are rules that in some measure set out to regulate use of a common resource, though the procedures cannot be reduced to an ecological economics. Rules encoded in textual traditions and folk and oral narratives teach devotees how to perform ablutions, how to think and act while on pilgrimage to Gaṅgā, how to enter sacred space, how to transport and store Gaṅgājal, and how to use it in pūjā and other rituals. Priests at places of pilgrimage today loosely "enforce" these prescriptions by preaching a similar set of instructions to pilgrims. They do not punish those who make minor transgressions; rather, they teach and direct.

The freedom from government control that religious practitioners hope to maintain today is, however, constantly threatened by official claims and now by official environmentalism. Contenders of the political-legal code argue that the state holds the right to regulate behavior by punishing transgressions of the law. Religious leaders, on the other hand, claim the moral right to teach and direct public behavior in the practice of ritual. But to enact their claims and rights, leaders at the Māgha Melā measure their respective powers in an imagined compromise with the state. In proposing such a compromise, the one set of rights grows out independently from the other. The result is that the government that holds the legal mandate to enforce a code of conduct loses moral or social power. Government officials buckle under the pressure of popular movements such as pilgrimages and rallies and are unable to exact the punishments they are in theory empowered to exact. Religious leaders, while teaching and directing behavior, have no political or economic power to enforce conformity to the code they speak and are reluctant to regulate the hands that feed them. In other words, religious and political leaders avoid coming together in the name of preventing pollution and create their own obstacles to obtaining the power they seek. Seeing Gaṅgā through different lenses, opposing groups create discursive distance by disregarding and even discrediting each other's traditions and bureaucracies. In its place, they exert what little specialized powers they have in their separate spheres, and protect these spheres from opponents. They eschew alignment with others in the view that alignment might undermine the powers they hold in their respective spheres. They are betting on strength in separateness rather than strength through collusion.

In 1987, T. N. Madan wrote that the separation of church and state in India could not be understood in terms of an American model, for there is no church to wall off, only the notion of neutrality or equidistance between the state and the identity of the people (755). Yet despite its inappropriateness, he lamented, the foreign model of separation has been adopted by some modern Indians, fostering a negative strategy toward religion in the modern nation-state. Madan argued that government and its secular ideology, by negating religion or attempting to remove religious power from public life, have made little headway in India. As Madan put it, secular ideology remains "encompassed" by the religious. While some scholars explain the cultural roots of this encompassment in order to raise a warning flag, others are more reticent and even welcome it as a sign of resistance to an orientalizing domination from the outside. No doubt, religious institutions, with their capability to teach and direct, are taking more significant strides in mobilizing the public, while government powers to regulate public behavior remain largely ineffective. But when defending Gaṅgā's purity, to protect their own territorial integrity religious leaders do not encompass secular and scientific ideologies. By pulling away from political leaders, they hope to prevent government intervention in religious affairs. But after pulling away and achieving that autonomous power, they do not attempt to reencompass political power by appropriating the messages of environmentalism. To remain powerful in their own right, religious leaders maintain claims to Gaṅgā in distinctly transcendent terms, distinguishing politics, science, and environmental pollution as pursuits of a separate order. Without appropriating the political message to "save" the river, they look to achieve their interest in autonomy-through-transcendence by ritual practices aimed at salvation from the material world.

Appreciation for Gaṅgā's immanent power, on the other hand, has not led to a motivation for environmental activism either. The presence of her power in this world does not mean that one has to act upon it or for it. That presence is everywhere yet apart from human agency. Gaṅgā is embedded in existence yet nothing can alter her. She is an eternal power acting in this world, inside, yet above the powers of humans and the ecologies they live in. Her forgiveness is limitless.

CHAPTER 11

Conclusion: Toward a Paradigm of Discourses and Powers

> Ethnography is actively situated between systems of meaning. It poses its questions at the boundaries of civilizations, cultures, classes, races, and genders. Ethnography decodes and recodes, telling the grounds of collective order and diversity, inclusion and exclusion. It describes processes of innovation and structuration, and is itself part of these processes.
>
> *(Clifford 1986:2–3)*

THIS TEXT HAS SUSTAINED AN OUTWARD LAYERING methodology to reach in and through multiple discursive configurations, bridging a consideration of discourses with a consideration of institutional rules, ritual practices, and flows of wastewater in a sacred river. The original inspiration for this study came from a discussion among tourists I overheard while boating along the river Gaṅgā. I have asked the reader to examine the historical roots of the tourist's vision of paradox in a critical way and to explore in an outward layering fashion the many other discourses about the river articulated by citizens of India. I have tried to show that the tourist's paradox is not a paradox at all for Hindus who believe in Gaṅgā's sacred purity. At the present time, the Hindu understanding of sacred purity does not render purity and uncleanness mutually exclusive categories and conditions.

A few of the approaches to the river Gaṅgā described here are tied to epistemic communities, to what Haas (1992) has described as the modern version of an imagined community. United by common ideas, beliefs, and values, epistemic communities handle the decision-making and policy-making affairs of corporations and nation-states. But I have argued that officials and scientists are not the only ones approaching the river as practiced, lived, and talked about universes; in fact, pilgrims, pilgrim priests, boatmen, and other pilgrim service providers engage in more intimate ways with Gaṅgā. Moreover, these policy-making, rule-defining epistemic communities appear rather ineffective at implementing the schemes they are set up to accomplish. Another set of intersecting discourses moves people's relationship with this river resource through different means and methods. In the process, what political scientists and international relations specialists consider to be the least epistemic becomes the most hegemonic; religious institutions are deeply involved with rivers, wastewater flows, and waste management policies and practices in many indirect yet powerful ways.

I have shown how the discourse on sacred purity is not simply an ideology that mystifies, or a symbolic chain that helps speakers and listeners misrecognize the true source and expression of their interests. It takes on these functions, but the hegemonic discourse and its substance and outcomes are much more, semantically, institutionally, and in terms of everyday practices. So as I move toward a discussion of how these varied discourses and powers might be recognized together, it is important to emphasize the cultural dynamic that makes the Hindu interpretation of Mother Gaṅgā so powerful. I have suggested that the resolution of serious wastewater problems in India today may have to come about through a resonance of these discourses, not through denial of the existence or significance of one or more of them. It may also have to come about through recognition that this powerful discourse provides cover for many abuses of the public good.

Following Foucault's general interest in excavating subjugated knowledges, I have outlined the confrontations between sacred and scientific knowledges, and the differences in practices and outcomes among ritual specialists, activists, and officials of waste management agencies. Yet some strategies appear to find a degree of continuity from colonial to postcolonial regimes. For instance, the strategy to undermine Hindu discourses on water resources has appeared several times over the course of modern history. In both colonial and postcolonial periods, the respective governments attempted to deny the validity of religious discourses on water resources, and in the process failed to achieve any hegemonic power. In colonial well water cases, justices tried to prove that caste impurity had no impact on potability or any links to identities that claimed privileges to scarce resources. Today, officials of the Ganga Project Directorate and the Jal Nigam continue that ambiguous relationship with the ecclesiastical realm. Their definitions of potability seek detachment from religious understandings, yet their everyday engagements with the public necessitate some lip service to these views and a significant amount of bending to religious practices and movements. At the discursive level, government officials deny sacred authority but at the level of practice they are powerless to quash it. The differences among Hindu, scientific-administrative, and activist discourses cannot be reconciled without grappling with these deeply rooted positions. The polyphony will likely remain.

While documenting the polyvalency of the terms *purity*, *impurity*, and *pollution*, I have suggested that the religious interpretation of Mother Gaṅgā claims the greatest measure of hegemonic power. I have shown that the semantic categories of purity, impurity, cleanness, and uncleanness are widely accepted by Hindu ritual specialists, pilgrims, and other residents of cities along the riverbanks. But I have tried to distinguish the notion of hegemonic power from other kinds of power circulating in and through the discourses and urban field sites highlighted here. In another way, government officials achieve a measure of power by negotiating bureaucratic and institutional rules and generating personal successes out of national "failures." This concluding chapter must confront the likelihood that this administrative form of power, something Ferguson (1990) likened to an "antipolitics machine," will remain entrenched in the near future.

In terms of the epistemic realm of policy, law, and the written constitution, the judiciary is making significant strides. For better or worse, these initiatives are bringing public policy debates into the courts, the media, and even city streets.[1] As public policy debates become court-driven exercises, the interchanges among judiciary, legislative, and executive branches work to shape Indian environmentality. But the goals and outcomes of court cases are highly controversial and often attenuated in their relation to each other, suggesting that agency resides with many other constituencies as well. Therefore, in this summary of forms of power, I must include the specific power of judicial activism at work.

Finally, these discourses are all about taming and managing river water and wastewater flows. They respond to the understanding that wastewater exerts many effects upon urban, religious, and everyday life. All citizens recognize that wastewater is a force they must come to understand and control, even if sacred purity transcends it. Wastewater flows run through the activities and cycles that integrate humans with water resources.

After summarizing how these discourses and forms of power intersect, I will conclude by commenting on the contributions anthropologists can make to the public policy interface. A collective movement against river pollution will not emerge on the basis of a single set of conceptual meanings for commonly used terms (chap. 2); but is it possible that a paradigm of differentiated voices can establish the parameters for a more inclusive discussion in the future?

Hegemony and the Discourse of Sacred Purity

The religious discourse on Mother Gaṅgā claims a hegemonic status in contemporary debates about wastewater flows and their many confluences with the Gaṅgā (see parts 1 and 2), taking *hegemony* as a form of consciousness achieved when citizens accept a particular set of moral, political, and social values as the natural order. However, this hegemony is partial, as Gramsci warned long ago; it is not always tied to ritual, administrative, or legal practices in a corresponding fashion. Breaks between this hegemonic discourse and everyday practices may occur in very routine ways. In order to summarize how it is that the Hindu discourse on sacred purity remains so important and so convincing to many citizens of India today, I must relate the key components of the discourse to the problem of essentialisms.

Since Orientalist scholarship has fallen under the purview of modern cultural critique, both native and foreign scholars working on South Asian problems have warned about the essentializing tendencies of historical and social-scientific writing and nationalist rhetoric and ideology.[2] Essentializing—the process of attributing fundamental essences to people, places, and movements in time—is a tool used by those employed in the business of representing others through discourse and the written word. It is a method the social scientist might inadvertently promote, since representations of self and other produced through cultural critique can be reproduced as essential elements of culture by other critics and the public. In concluding this text, I must explain how these essentialisms help citizens achieve hegemonic power for the Hindu discourse on sacred purity.

The resistance power residing in the discourse on sacred purity is deeply associated with the discourse on motherhood. Gaṅgā is a mother, preserver, sustainer, and purifier. In the midst of the worst threats to humankind, she perseveres without becoming defiled herself. With the continuous strength of śakti, the female form of power, she forgives humans for their attempts to exploit her (which are never fully successful) and for the gandagī they bring to her. She never shies away from her duty to clean up the world. Her power is greater than the powers of those who use and attempt to manipulate her, evidenced by the sunken temple in Banaras and the Ganga Action Plan. This motherly essence functions on two levels. For Hindu pilgrim priests, it allows paṇḍās to rule out other visions of her that might undermine the power of their own professional duties. Sacred purity rules out the possibility of impurity, though it does not rule out uncleanness. In transcendence, purity cannot be attacked; but in this world, uncleanness proliferates. For paṇḍās and other religious leaders, Gaṅgā's purity is transcendent as it endures the degeneration of the Kalī yuga. Uncleanness is deposited or laid down on, but does not affect, the sacred layer of purity. By this reasoning, paṇḍās maintain their occupational specializations amid the degenerative processes of Kalī yuga.

By emphasizing purity over fertility, paṇḍās and other religious leaders also valorize Gaṅgā's cleanup skills rather than her reproductive power and associate the former with the values of motherhood in general. Gaṅgā and all mothers exhibit the same features: selflessness, cleanup work, and forgiveness. This valorization of purity over fertility is one that is linked to the male bias in the testimonies I collected, since men appear to know more about purity–because it is connected to their own occupational activities–and less about fertility, a power that in some respects gives women a measure of security and independence vis-à-vis men.[3]

In a way unintended by those who believe in sacred purity, Gaṅgā is used in more aggressive ways for agricultural production, for the generation of hydroelectric power, and for the carriage of domestic and industrial waste. Yet unlike her believers, these resource users expect that she can clean herself through aeration and the augmentation of flow, not through sacred power. Industrialists and government officials use the citizens' valorization of motherliness as a screen for their own exploitation. Behind this screen, they enjoy many liberties in dumping practices and engage in many a breach of orders and laws.

More broadly, the essentialism of Mother undermines the efficacy of scientific and governmental statements about pollution and weakens activist discourses on environmentalism. In the colonial period, laws and policies on water resources did not achieve hegemonic power. Colonial officials attempted to replant British notions of the ecclesiastical and the civil in the soil of Indian villages and use them to reorder the uses of certain freshwater resources. But legal cases on well water conflicts have shown that despite its interest in hegemony, the colonial government pursued the goal of administration in a rather contradictory way. Though achieving a legal literature of some substance, they did not usher in a new set of acceptable meanings for water resources. Well water cases are the first evidence of governmentality in the field of water and sewage management, even though they were not successful in convincing the public to see water resource use in the same way. Instead, cases on well water disputes developed notions of defilement, fouling, and pollution that were revised later to make new administrative rules and underwrite the concern for potability. Today, postcolonial religious leaders reproduce the sense of the separate domains of the civil and ecclesiastical that colonialists established, but they do so to keep official interests at bay. My data show that the postcolonial/religious notion of separation, rather than the colonial notion of the separation of domains or the postcolonial/official version of separation, has had more significant effects in terms of hegemony. That is, by and large, the citizens of India I interviewed consider Hindu discourses on Mother Gaṅgā's purity to be more convincing than postcolonial definitions of pollution. They appreciate rituals such as ablution and āratī more than secular calls to distance everyday life from the river in the name of environmental pollution. The postcolonial/religious version draws its persuasive power today from religious symbols and ritual practices. Current calls for a separation of domains by Hindu religious leaders turn the colonial notion of separation against official control and open up a space for ecclesiastical hegemony. These postcolonial/religious domains are tied directly to the essentialism of Mother Gaṅgā.

Today, the Hindu view idealizes Gaṅgā as a mother goddess with the power to provide for humanity without reacting to the ways humans use and worship her.[4] In her transcendence, Gaṅgā is without reactive power; her sacredness is idealized as a fixed state. In her immanence, she flows through dry and monsoon seasons, drawing upon her sacrality to clean up impurity and uncleanness. As religious leaders attempt to separate the domains of the ecclesiastical and the civil, they maintain a contrast between scientific/secular understandings of the Gaṅgā and Hindu/sacred understandings, and they allow that contrast to generate tensions between theories that put humans above "nature" and those that render Gaṅgā a transcendent, divine force in human existence.[5] But it is important to remember what is being separated. As outlined here, there is no separate domain of "nature" in the Hindu discourse. Instead, Gaṅgā, a deity, subjects (in very humble and loving ways) both humans and ecological processes to her own sacred design. By embodying and constituting ecology, she allows humans access to it. In this view, there is no nature that is separate from humans. Both are tied together in Gaṅgā's enveloping power.

The scientific-administrative discourse defines Gaṅgā as a component of a river ecosystem. The river undergoes constant changes in physicochemical properties and flow regimes. The Gaṅgā is affected by diversions and obstructions of flow, by the disposal of waste and wastewater, and by ecological changes in its watersheds. The river alters course, moves mountains of sediment, floods human settlements, and flushes bacteria and toxins, within the contours of ecosystem processes. But scientists and government officials tend to demarcate this ecosystem in rather fragmented ways, drawing boundaries according to the visions of particular agencies and academic units that fund and carry out scientific research. Specific scientific understandings of the river are generally limited to particular reaches or to particular relationships between a limited set of ecological variables. Moreover, the scientific-administrative vision of the river assumes a continued growth in human settlements and consumptive and market demands. Most scientists and officials explain that as human demands for water resources increase, the river will have to adapt.

Many of the male environmental activists I have introduced tend to see the river as both active and passive. They essentialize the river as a changing resource affected by human activities; the river is affected by urban growth, industrial production, and the practices of citizens and administrative and industrial elites. Activists believe that Gaṅgā requires advocates who can save her and save humanity at the same time. They hold a resistance advocacy, one that attempts to take the power of resource use out of the hands of administrative and industrial elites and offer it up to the vast public interest. The administrative and industrial elites against whom they resist are clearly defined, while the vast public interest remains inchoate. Saving and cleaning the river are emancipatory exercises that work upon a Gaṅgā needy for advocacy.

The gender equation, while clearly evident in the Hindu discourse, is almost absent from the scientific-administrative and activist versions. More often, the latter two are sustained *against* a religious view of gendered ecology. The positions of government officials and scientists are, in many cases, conjured up in contradistinction to caste or religious views and sometimes made to indirectly deny their existence as they rename the river in ecological, this-worldly terms. Most but not all environmental activists also deny the river any transcendent power, as they search for engineered, legal, and administrative solutions to river pollution problems. The Mother Goddess is denied and downplayed in their interpretations as well. The agency that these approaches attribute to ecological processes is not connected to the agency of the female or motherly spirit. Rather, it is an agency that is essentialized as a neutered ecology, without any specific association with the power of Mother.

With Mother and without Mother, the three visions—the Hindu, the scientific-administrative, and the activist—are different in terms of their semantic structures. Gender works for one and not for the other two. Yet it works so powerfully for the Hindu view that it forecloses most of the communicative potential that might exist between that view and the official and activist views. Denial of Gaṅgā's gendered, transcendent essence, of her power as a separate domain, and of the separate supervisory role of religious leaders and ritual practitioners mutes the persuasiveness of official and activist logics.

Negotiating Rules and Disjunctures

This analysis of the relations among discourses is helpful for understanding how specific logical and symbolic components become persuasive over time. But discourses are not the only means, or vehicle for, power. Where power resides in the production of goods and services and in the bureaucracies that monitor and regulate this production, we find another set of strategies that operates through administrative and industrial practices. I have outlined these strategies by breaking down environmental governmentality into its historical components and relating secular discourses on pollution to other dominant discourses over time. Administrative power occurs as officials negotiate a series of explicit and implicit bureaucratic, institutional, and legal rules and disjunctures. Government officials may manipulate these rules and disjunctures to generate personal gains. So in this case, governmentality is not about generating a viable hegemony, a persuasive code that the public subscribes to; it is about using discursive positions to master layers of rules and institutional disjunctures and negotiate a way through them.

Chapters 8 and 9 explained the ways in which government officials, industrialists, and an amicus curiae developed relations of support and benefited from a legal case filed by an environmental activist. Government offices were blamed for dereliction of duty, and industries were held accountable for noncompliance with court orders. But in the midst of these legal discourses, administrative elites negotiated levels of rules, breaking some and claiming to adhere to others. They juggled rules and worked through institutional disjunctures to generate personal achievements. As a consequence, industrialists continued to pollute.

These movements through layers of rules (that sometimes require jumping from one semantic plane to another) constitute a particularly successful form of administrative power. Rules and institutional policies are negotiated at global, national, and city levels. There are several layers of rules and policies at the global level alone, produced by the Dutch government, the Indian government, and international laws and policies. The Dutch government and its many mediating agencies make an agreement with the government of India to provide a certain amount of grant money to carry out specific kinds of waste management and institutional development tasks in Kanpur. The agreement includes objectives that lay out the procedure for distributing the grant and executing planned work. However, these objectives operate almost wholly within the national domain and, on an everyday basis, at the local levels where projects are implemented. The Dutch government exercises little control over the actual outcome of domestic institutional arrangements (see chaps. 8 and 9). So while money and technical assistance and consultancy are passed from global to national levels, their translation and use are largely domestic affairs. In the past, World Bank officials effected changes in water and sewage policy and practice in municipal, state, and central departments. They did this when breaking the Local Self-Development Agency into the Jal Sansthan and the Nagar Nigam in 1974. But the projects carried out under the Ganga Action Plan today reveal more domestic control of environmental programs, even while much of the financing continues to come from foreign sources. This has created problems for foreign donors who expect to exert more domestic influence when promoting their technological agendas.

In chapters 8 and 9 I also traced how money, professional contracts, and responsibilities run between central, state, and city government levels within the nation-state. Money flows from the center to the local level with significant control held at the central and state levels and very little at the city level. A paltry amount of revenue is generated at the city level for local uses, with almost nothing percolating up to the state or central levels. In addition, professional contracts move from the state level to the city level and back again. The states hold most of the power to decide which agencies will get the contracts to do the work under a specific funded project. City administrators simply oversee the contract as it is implemented by the professionals chosen by state officials. Finally, officials in the highest reaches of the state and central governments make appointments to key posts in the Jal Sansthan and Jal Nigam, and they put high premiums on those plums. Yet state and central government officials often transfer these local-level affiliates at the whim of politics, creating a shifty realm of bureaucratic politics at all levels. These are examples of the policies, rules, and institutional disjunctures that administrative agents negotiate to generate personal successes.

Legal against Antilegal Activism

Public interest litigators have attempted to put a stop to the practice of generating personal successes out of national failures. Legal discourses often work for national and global development; but lawyers and citizen petitioners also attempt an active resistance from within the constitutional framework. They can invoke several provisions in the Indian Constitution to activate legal reform. In public interest cases, justices may, if they wish, turn these citizen invocations into judicial activism.[6] This gives activist discourses, when articulated through the legal process, a transformative power. However, serious problems in compliance exist; legal discourses are not always able to achieve cultural and political transformations. Instead they may facilitate the administrative form of power. In some instances in the Gaṅgā pollution cases, court orders became boons for government officials and playthings for industrial agents.

In a more general way, these legal cases show that noncompliance is linked to the privatization of state bureaucracy, to the rogue behaviors of industrialists, the capitalist elite, and police agencies, *and* to the circular traps in which citizens find themselves caught. Disjunctive institutional rules also make the public confused about how to hold official institutions accountable for the responsibilities they are legally required to perform. Meanwhile, development-capitalist actors learn how to slip their agendas through these official fractures and public confusions. Officials of the Uttar Pradesh Pollution Control Board, for example, are conscious of the implications of their practices because they draw personal successes from them and are reluctant to make the structural changes necessary to eradicate the conditions for these successes. The public, while aware of their practices and the motivations for them, becomes confused about what is actually happening or is frightened into silence.

Despite its limited hegemonic and administrative accomplishments, public interest litigation constitutes an active center of resistance and a counterdiscourse to development for some Indian residents.[7] This resistance is not found in a movement per se, but in legal debates about rules, policies, and the written Constitution. Upendra Baxi has argued that judicial activism is a response to the problem of the lawlessness of the state. He explains:

> At the theoretical level of analysis of state, it has now been established that the activity, role and place of the state stretch a very long way beyond law and judicial regulation. Indeed there is no way to overcome the fact that the state is the foremost violator of the law as any reader of Indian Law reports can make out fairly quickly. The law itself provides for the manner of its effective, redressless breach on many an occasion in India. (1985:15)

Legality shot through with illegality, as Baxi puts it, is used as a resource for political legitimation when dominant classes expose illegality as a way to reinforce the legitimacy of their own regime.[8] But citizens, he adds, can also use governmental lawlessness to delegitimate the dominant classes through judicial activism. It is at this point that lawlessness becomes a counterdiscourse. Exposing the lawlessness of the state transforms the state's resource into a liability (Baxi 1985:16). This is often linked, as my data sets have shown, to a struggle for the deprivatization of the state. The model of the *written* Constitution of India is used in the counterdiscourse that assails the *unwritten* Constitution of India (Baxi 1985:18).

Therefore, some citizens continue to see public interest litigation as an essentially futile exercise, because, as the data have demonstrated, there is a tendency for that litigation to be appropriated by official powers and leaders of the executive midstream during or after orders and directions have been made by the court. The struggle is never completely victorious, which leaves many citizens with a feeling of helplessness. The other side of this public cynicism is linked to the understanding that Brass has alluded to, that many Indian citizens are already litigious and know the ways that lawlessness can be masked or used as a counterargument. It is difficult to find the truth in it or to find a way to enforce the "written Constitution." This is what makes the discourse on sacred transcendence so attractive. In the unending circularity of corruption, denial, and lawlessness, there is some hope. That hope, for many Hindus, lies in the immanent and transcendent. The *current* power of this discourse about Mother Gaṅgā is directly related to the character of the state in India and to the character of the public's relationship with the state. The only way to counter state character, as Swami Chidananda has warned, is to completely reject it, to proffer a world of purity beyond it, and to have faith that transcendent purity will ultimately neutralize worldly dirtiness and greed. It is thus the way with wastescapes. They embody all that is degenerative in this world, and all that worldly action produces. This is why wastescapes cannot be cleaned up by the same degenerative powers that created them. Finally, in the midst of these domains of the ecclesiastical and the civil, the sacred and the environmental, we find the industrial elite in a cozy silence, hiding behind official forms of environmentalism and notions of Mother Gaṅgā and sending only their best lawyers to answer petitioners and justices in the country's courts.

The Power of Wastescapes

Looking into hegemony, administrative power, and the transformative potential of public interest litigation, we come to the central role of wastewater flows. These flows, also called wastescapes, are pervasive and pernicious and should be made a visible part of models of human-environment interaction. They represent the gross outcomes of industrial, urban, agricultural, and household practices. Wastescapes meander at the margins and interstices of everyday life, yet are the direct results of all the consumption and production processes so central to human existence. In the Gaṅgā basin, wastewater flows originate at the level of the locality and then diffuse outward. After drawing water from river and groundwater sources, members of households, municipalities, agricultural farms, and industries produce wastewater. This wastewater flows into drains or nālās, streams or rivers, circulating through surface water and then percolating into groundwater. When joined with other water streams, wastewater circulates through local river ecosystems and beyond them to waters that flow transregionally.

Following Rees and Wackernagel (1994), I have shown that the ecological footprint of urbanism is deep when cities are located along streams and rivers.[9] The Central Pollution Control Board reports that three-fourths of the pollution of the river Gaṅgā comes from the discharge of untreated municipal sewage in Class I cities (cities with populations above 100,000). This wastewater diffuses out beyond households and industrial production units through a maze of nālās and eventually beyond the city by way of the river's flow. Moreover, as industrial practices increase in urban centers, effluent levels rise.

Wastescapes will continue to serve as raw data for the industrial practices a researcher cannot witness from the inside. They are a powerful metaphor for talking about industry in the context of environment and culture. There is no denying them and often no way of hiding them for very long.

The power of wastewater lies in the fact that it continues to elude or escape the many structures that attempt to define and manage it. At the ethnosemantic level, there is no agreement among the public about the very epistemology of the confluences of wastewater and sacred rivers. There is a great deal of disagreement over reports on water quality, discharge, and flow levels. As wastewater flows through the interstices of everyday life, it threatens to disrupt human and other forms of life over the short and long term. In built environments, wastewater meanders through the disjunctures between gravity and electrically based wastewater diversion and treatment systems, weakening urban construction and frustrating attempts to divert and treat it.

Intersections and the Public Policy Interface

Ethnography has shown how the terms *purity*, *impurity*, *cleanness*, *uncleanness*, and *pollution* operate as centers of discussion for various interest groups. I have called these centers of discussion *intersections* to bring to light the multilayered nature of communication. These are intersections that start with concepts and words and then point our attention to other practices and flows. On the other hand, these intersections are distinguished from the centers of resistance forming within each individual discourse. For example, the Hindu discourse on Gaṅgā's purity provides a center of resistance to development, science, and official environmentalism, but also offers up points where it intersects with scientific-administrative and activist discourses. Within the discourse on purity, Hindu devotees talk about uncleanness and thereby interface with administrative discussions of environmental pollution, as long as pollution is not used to deny the existence of sacred purity. The Hindu discourse also valorizes flow and comes close to scientific-administrative and activist understandings of the need for flow and aeration to degrade wastewater. Moving in other cultural directions, the term *purity* also connects Gaṅgā's sacred power to survival and chauvinism, the occupations of men, and rituals of birth, death, and celebration.

The job of connecting Gaṅgā's transcendent purity with the necessity for active worldly reform is a complicated and underperformed task. To date, religious and political leaders have not been able to join the worship of the transcendent with an active interest in protecting Gaṅgā's immanent cleanness and physical form. My own experiments in connecting these transcendent values with cleanup projects have failed; but they have convinced me that religious leaders are well positioned to make a contribution. Understanding devotees' beliefs, they can bring together worship for Gaṅgā's transcendent purity and concern for her immanent role as Mother. There are several religious leaders who have begun this experiment in their own ways, and I watch their work with great curiosity. On the other hand, potential exists in the legal pathway, so long as legal interventions are open to public debate and allowed to achieve their goals for compliance.

In conclusion, I emphasize that ethnographic writing can play an active role in constructing a paradigm for public debate. The ethnographer can "act" in this discursive field by explaining this polyphony and making it part of an explicit paradigm, part of a recognized set of parameters for public debate on water and river concerns. Ethnography may also be joined with other forms of applied work or academic activism. During the latter part of the research and writing of this text, I worked with an Indian colleague to contribute to policymaking on pollution control and wastewater management in India. Together, we arranged for a series of meetings among Indian and American lawyers, scientists, and members of nongovernmental groups to discuss river pollution and wastewater management problems. Our original idea was to bring the Indian voices together on a common platform, and take the help of American professionals to establish a more permanent resonance of discourses in the realms of policy-making and legal debate. Although the Americans did not promote any specific technological agenda, they did promote the interest in making these differentiated voices understandable over time and helped to argue for their inclusion in a publicly declared range of concerns. This is the kind of contribution that a nonnative anthropologist can make to a public policy debate in a foreign country. In this case, however, the real work of sustaining these discourses as part of a public paradigm, of maintaining them as part of an accepted range of visions on common problems, will be done by the various interest groups outlined in this text.

As an American analyst, I have entered into debates over Gaṅgā's purity and cleanness, her flow, and wastewater management in urban riparian zones. Looking more critically at my own contributions, I must admit that I was not just hoping to promote a democratic reckoning of positions. I was, in effect, making value judgments and debating definitions of the public good—helping to institutionalize certain priorities and theories. To be sure, the implications of my Americanness have been vast and problematic. American "concern" for the Gaṅgā, like American environmentalism in general, appears contradictory and shallow to those critical of the links between environmental managerialism and globalization today.

There are also implications for my particular choice of heroes. Clearly I give members of several NGOs the higher ground and members of some government agencies a shake of the finger. As members of a few NGOs became acquaintances and friends through research and our applied project, they had to face the consequences of that interaction. Other Indian citizens began to link them to American environmentalism and to the controversial processes of globalization under way. It is necessary then that anthropologists be cognizant of the ways ethnography renders certain subjects congenial and then implicates them in the author's vision or analysis. It is the anthropologist's job to examine that rendering as part of the problematic being investigated. I have tried to do that in a broader way by beginning with a Western tourist's paradox and then moving beyond that vision of paradox to the many other concerns and practices that constitute interactions with the river Gaṅgā. The tourist's paradox, rooted in concerns familiar to my own culture, was dismantled and then recoded as part of a larger debate. The outward layering methodology provided a way to disentangle various approaches to the Gaṅgā, delineate distinctive discourses, and excavate their intersections over time. This method might help others to understand the various positions emerging in debates over environmental problems across the globe.

Notes

INTRODUCTION

1. I found most of these materials in the library of the Indian Law Institute and the Jawaharlal Nehru Memorial Library in New Delhi.

2. Quite early in my research, I found myself situated in a domestic debate, a debate that made two oppositions immediately apparent. The first opposition highlighted the polarities between common residents and the government, and the second pointed to tensions between East and West in globalization and development. Over time, these oppositions proved powerful enough to shape my own identity and data collection activities. At first, I assumed the role of Western environmentalist promoting a commitment to clean up the earth. This is a modality that also fits foreign tourists who look for clean places in which to discover the exotic other (Alley 1990; Argryous 1997:169). But, alas, my proselytizing was quite unsuccessful; I rarely got others to debate environmental issues in my terms! The few that did appeared to find different purposes for their activities and eventually forced me to approach the discussion by way of other cultural considerations. That is when my own learning began.

During the 1990s, foreign investment and international aid and loans to private and public sector companies generated a substantial rise in cash flow throughout the country, causing the United States to dub India a "big emerging market." In the midst of this, foreign observers continue to appear suspicious to Indians. In particular, Indians trained in the ideology of nonalignment are not won over by the newfound enthusiasm for trade development and growth. This position was evident when some members of nongovernmental organizations asked why I was studying river pollution in India and not in my home country. But these critical observers also turned this suspicion inward, toward their own bureaucratic and development structures and toward the variants of official environmentalism that seemed to develop alongside them.

3. When informants started to become close friends or even fictitious relatives of mine, then the work, in my opinion, got more difficult. In the Hindu sacred city of Banaras, I developed strong relationships with both pilgrim priests *(paṇḍās)* and boatmen. This became a source of conflict in *their* minds, since boatmen see paṇḍās as exploiters, and paṇḍās see boatmen as unqualified informants. Since my (fic-

titious) brother is a boatman, some paṇḍās found that I crossed caste boundaries in awkward ways. On the other hand, I was warned by boatmen on several occasions not to trust paṇḍās. Even though these warnings did not restrict my movement, the discontent between these two groups brought to my attention the challenges faced by environmental activists who attempt to unite people of different castes under new symbols. Although I was never fully able to resolve the tension between these groups because it lies at the foundation of social hierarchy on the ghāṭs, both groups of informants eventually accepted my ambiguous position.

4. As Lynch (1990:18) describes it, the ideal is a striving toward understanding by drawing out the divine bliss inherent in all humans. He has noted that in this ideal there is no emphasis on identifying emotions with the lower self or on recasting the search for realization as a quest for individualism. Emotions are, quite differently, one's true self and the essence of true reality. Lynch and others have argued that anthropological discourse often pushes emotion to a subordinate position, in order to privilege rational structures of cultural existence. See also Abu-Lughod 1986; Lutz and White 1986; M. Rosaldo 1980, 1984; R. Rosaldo 1989.

5. *Ji* is an honorific marker that is added to names of gods, goddesses, and persons.

6. Baviskar (1999) and Brosius (1999a) have pointed out that environmental anthropology has overlooked urban-industrial topologies, in favor of studies of "natural" rural landscapes. This is an important issue to raise and one that I take very seriously in this book. Urban contexts are settings in which some of the most serious cases of river water contamination and pollution occur in India and in other parts of the world. They must become central to the anthropological investigation of environmentalism and ecological change.

7. This awareness of a network of affected citizens is evident in the commentary made by a prominent sādhu-sant about his antipollution rallies. He said, "On the 31st of October we met with the governor of Bihar and demanded that necessary steps be taken to treat dangerous industrial waste and the waste of cities along the river, to give extended assistance to the need to sustain marine life and fish in the Gaṅgā for fishermen and boatmen, save the cutting of farm land in neighboring villages and make the Gaṅgā free from pollution in Bihar." In this statement he called upon a set of concerned players—fishermen, boatmen, farmers, and other residents of cities on the riverbank—to form a common bond in their concern for the river.

CHAPTER 1

1. Tsing (1993) has also combined the Foucauldian notion of discourse with Gramsci's notion of hegemony to understand culture and power in the relations the Meratus of Indonesia forged with the state. As she put it, "the former vocabulary [that of Foucauldian discourse] is particularly useful for bringing together issues of meaning and practice in examining the construction of power; the latter [Gramsci's notion of hegemony] calls attention to divergent political stakes in processes of social transformation" (1993:303, n. 1). Although Gramsci's work dealt with a different time and culture, his notion of hegemony is timeless and not culture-specific in that it measures power as the extent to which intellectual and moral leaders or cultural groups are able to persuade a wider group to agree to

their moral, political, and cultural values as the natural order. In Gramsci's sense then, power is measured in terms of public demonstrations of consent, in the extent to which individuals follow a cultural pattern and articulate certain assessments and worldviews.

2. This is what others have called *ethnoecology* or *traditional ecological knowledge.*

3. Ethnographies by Lansing (1991) and Peluso (1992) are especially noteworthy.

4. Kottak (1999:26) made this kind of characterization to explain the ways that traditional or local ethnoecologies are challenged, transformed, or replaced in contemporary global interactions. He refered to developmentalism and environmentalism as Euro-American ethnoecologies that spread into myriad cultural settings, influenced in the process by national, regional, and local ethnoecologies. See also Kottak and Costa 1993; Costa et al. 1994.

5. Brosius (1999a:281) has pointed out that the emphasis on contestation in environmental anthropology is becoming increasingly formulaic, having the effect of routinizing and naturalizing contestation. While I take contestation as a starting point for discussion, my aim is to make the specific form of this contestation more apparent through analyses of linguistic particularities and shifts in semantics over time.

6. Although *pradușan* is linguistically close to *pradușit*, which means fully spoiled, it is meant in local discourse as a transliteration of *pollution.* As the word *pollution* developed in academic, governmental, and legal discourses of the late twentieth century, many who became familiar with these discourses learned the term in association with the institutions that executed actions against pollution.

7. Here I am thinking of Kenneth Burke's work (1955, 1973) on rhetoric and metaphor as key elements of social solidarity. Metaphors help people to identify with a common symbol and thus unify an otherwise disorderly or dispersed group of individuals. Fernandez (1974) has also developed the idea that metaphors hold transformative power. Metaphors start at a familiar point and then move people to identify with an unknown common denominator.

8. Studies of the cultural construction of risk extending from the work of Douglas and Wildavsky (1982) have focused largely on issues of nuclear power, with fewer scholars applying their model to the analysis of waste politics. See, for example, Downey 1986; Gerlach 1987; Gerlach and Rayner 1988.

9. More recent surveys of environmental values conducted by anthropologists have also neglected to ask informants about this area of concern. See Kempton et al. 1995.

10. See Curtis 1992; Cvetkovich and Earle 1992; Dake 1992; Gerlach and Meiller 1987; Hallman and Wandersman 1992; Kasperson, Golding, and Tuler 1992.

11. Beder has numerous publications to her credit (see bibliography), many written for a broad public audience.

12. See Buttel 1987, 1992; Buttel and Flinn 1974; Buttel and Taylor 1992; Schnaiberg 1973, 1980, 1985; Stallings 1973; Turner and Killian 1972; Walsh and Warland 1983; Wood 1982; Zald and McCarthy 1979.

13. Alley et al. (1996) have offered a different reading of the transformations undertaken by an environmental grassroots group in the United States, arguing

that a local group can disintegrate once its members begin to link their concerns with a national environmental organization. In the case we highlight, linkages with national organizations ultimately led to the demise of the local group.

14. The field of environmental justice is a newer focus and includes studies of race, justice, and environment from various approaches. For examples, see Bailey and Faupel 1992a, 1992b; Bailey et al. 1993; Bryant and Mohai 1992; Bullard 1983, 1992, 1993; Checker United Church of Christ 1987; Grossman 1993; Miller 1993; Novotny 2000; Szasz 1994.

15. See Bryant and Bailey 1997; Hershkovitz 1993; Peet and Watts 1996; Peluso 1992; Picchi 1995; Schmink and Wood 1987; Stonich 1993.

16. Eder (1996) has made a similar claim, that the environmental movement no longer dominates the discourse on the environment. There are now many voices that claim a stake in the environment, and this plurality is transforming the public sphere. According to Eder, ecology now provides the master frame for the perception and experience of social reality. Using frame and discourse analysis to make his argument, Eder focuses primarily on contexts in which actors articulate naturalist notions of ecology; these are contexts that the reader assumes exist in the Western world. While I agree with Eder's fundamental position on the dispersion of meanings of environmentalism, I consider his model too rigid for use in the context I am exploring. Cultural constructs associated with environmental discourses in India are even more pluralistic than he suggests, and require that we open up our methodology to explore symbolic expressions of a wider range.

17. Rangan underscores the emergent nature of development in India in her description of grassroots struggles in the Garhwal Himalayas. She argues that social protests in postindependence India are concerned with access to development and with forcing the state to assume greater responsibility for addressing problems of uneven regional development and social equity (1996:206). New social movements, she argues, are not against but part of the idea of development. Looking at interest groups in waste politics, I agree that development is envisioned as a great light on the horizon, not yet fully captured and realized. But I would also argue that the various positions taken by these interest groups indicate a more critical array of arguments about development and nationalism. While development may be a fuzzy goal for many, there are some who bring up serious concerns about the fundamental sacrifices it entails. Moreover, a focus on current dynamics in waste-related discourses and development practices needs to look at the connections between resistances to postcolonialism and resistances to the waste byproducts of modern capitalism (see Nandy 1983, 1995).

18. Sharma (1998) describes an international climate-change conference in Argentina in which American participants were clearly attempting to dominate the discourse to avoid committing to the Kyoto Protocol. Sharma argues that the aim of U.S. representatives was to get developing countries to commit to voluntary reductions before ratifying any treaty to reduce emissions at home and that they clearly favored emissions trading over cutting domestic emissions as a way to meet commitments under the Kyoto Protocol. More recently, Anil Agarwal has made this kind of critique in the December e-mail newsletter of the Centre for Science and Environment in New Delhi at ⟨http://www.oneworld.org/cse⟩ (Dec. 20,

2001). Gupta (1998:295–96, 311–13, 321–26) has also discussed the positions taken by American leaders in climate-change debates and Indian farmers' responses to these international conferences and the treaties produced by them. See also Eder 1996 for his view of the rhetorical angles on environmental conservation pervading the public sphere today.

19. The multiplicity in this scenario is similar to the diversity Orlove (1991) found in the region of Lake Titicaca, Peru. In contests over lake resources, a lack of communication between groups competing for control of the same natural resource led to a major disjunction. Each group imagined that the other had accepted rather than contested its position, as members participated in multiple systems of resource use. Unlike Orlove, I am arguing that this multiplicity does have intersecting centers and that they function more to obscure the ways officials operate through institutional disjunctures.

20. This differentiates popular debates about river water in the Gaṅgā basin from water debates in the American West. While the former are more about quality, the latter center on quantity. In the American West, permits to ownership of river flows were established during the era of settlement and mining, and they continue to exist today as part of a centralized system of state and federal control. They are considered by many analysts to be the source of many of the West's water problems. Though couched in the idiom of scarcity, water problems have much more to do with this system of distributing use rights and with the legacy that system has established in water allocation today. Based on seniority or prior appropriation, use rights favor those groups (primarily agriculturalists) who acquired permits to rights over a century ago when the American West was shaped by very different political and economic agendas (see Reisner 1986, 1990).

21. Measuring the extent of dumping and physicochemical pollution borne by an industry is beyond my own expertise, so I relied upon the studies of others and had to decide whose data were accurate and reliable. Industrial interests and uses are introduced but discussed in terms of how members of wider debates perceive and contextualize them.

22. Few in the common-resource literature have applied this focus on the roles of nongovernmental or unofficial players in resource management to the analysis of river basin resource users. A recent book and article on water and power suggests, however, that this focus on rivers may soon come of age (Donahue and Johnston 1998; Williams 2001).

23. This may be due in part to a projection of the dominant concerns of U.S. water specialists who have tended to privilege quantity over quality. Recent TMDL (total maximum daily load) legislation operating in forty states attempts to shift the focus more toward quality, even though increases in quantity continue to be seen as a viable antidote to pollution or degraded water quality (Adler 1999).

24. Here I am using Gramsci's notion of hegemony and considering power as the ability to create conformity to a vision or model that is worked out conceptually and semantically (see also chap. 1, n. 1).

25. See Paul's (1997) reference to belling the cat. Belling the cat is a metaphor for warning of danger or exposing injustices such as corruption or mismanagement in government projects. The story goes that a group of mice decided to tie a

bell around the neck of a cat that threatened them, so that the bell would warn when the cat was near. The mice, however, could not decide who among them would actually creep up to the cat and tie the bell around its neck. No mouse was ready to risk its life to tie the bell. Likewise, many activists admit that it is a risk to their lives to expose the injustices perpetrated by the elite. In Hindi, the phrase is: *Bili ke gullie par kaun gunti bandegi?* (Who will tie the bell on the cat's neck?)

CHAPTER 2

1. Quoted in a photo exhibit at Magha Mela, 1996; and partially reproduced in Ministry of Environment and Forests (1994:8).

2. See National River Conservation Directorate web page: ⟨http://envfor.nic.in/nrcd/nrcd.html⟩ (access date 12/20/01).

3. See Marglin 1985a; Babb 1975; Das 1977; Feldhaus 1995; Fuller 1992; Raheja 1988.

4. See Alter 1992; Daniel 1984; Inden 1976; Marriott 1989; Marriott and Inden 1977.

5. He drew upon the distinction between culture and nature to argue that the Hindu ethical universe privileges heredity but does not see the threat to culture that might be posed by an "irruption of the biological." Apparently Dumont distinguished between purity and the unfastidious, and suggested that a collapse of this distinction would bring about the naturalization of the elite he belonged to.

6. See Callicott 1994; Capra 1991; Chapple 1998; Chapple and Tucker 2000; Hargrove 1986; India International Centre 1992; Lovelock 1988; Singh 1993b; Tucker and Grim 2001.

CHAPTER 3

1. With a catchment area of 97.6 million hectares, the surface water potential of this system has been estimated at 55.03 million hectare-meter (Bhargava 1981b:142).

2. "A River of Death? Scientists Detect Cancer-Causing Agent in Yamuna," *Asian Age*, 22 September 1998, 2, 11.

3. See Dimmit and van Buitenen 1978; Eck 1982b; Kinsley 1987; O'Flaherty 1981:231–32; Sivaramamurti 1976; van Buitenen 1973:159, 216–20.

4. *Rig Veda* 10.9 and 7.49; translated in O'Flaherty 1981:231–32. See also Chaubey 1970.

5. *Mahabharata* 1(7)64.20 and 1(7)91.1–93.10; quoted in van Buitenen 1973:159, 216–20.

6. Das Gupta 1984:39.

7. Das Gupta 1981:27.

8. "The Spoiling of the Ganges," *Hinduism Today*, October 1997, 22.

9. See van der Veer 1987:283–301, 1994, 1996:254–77.

10. In Madya Pradesh the BJP ruled from 1990 to 1992. In Uttar Pradesh they held the government from 1991 to 1992. In 1995, the Bahujan Samaj Party (BSP) split with the Samajwadi Party (SP) with BJP support, but it really was a BSP government. In 1996, the BJP formed another coalition after elections with the

BSP. Then the BJP split from the BSP in order to head the government. In Rajastan, they held power from 1990 to 1992 and from 1993 to 1998. In Maharashtra, they were in power from 1995 through the writing of this book in early 2000. The governments that ended abruptly in 1992 had been imposed after the demolition of the Babri Masjid.

11. The events in Ayodhya in 1992 brought a great deal of attention to bear on the issues of religious nationalism, secularism, and communalism. For discussions of these issues, see Jha 1995; Madan 1993:667–97; Miller 1991:783–92; Mitra 1991:775–77; Parikh 1993:673–84; Subrahmaniam 1995; Thakur 1993:645–64; Upadhyaya 1992:815–53; van der Veer 1996. Comments by T. N. Madan (1987) several years earlier make it clear that secularism has been a conundrum for scholars and nationalists since the India's independence and Nehru's first designs for national policy.

CHAPTER 4

1. Saussure (1966) defined the sign as a two-sided psychological entity, composed of the sound-image (signifier) and concept (signified). He posited that the two sides are intimately united and recall each other. I use Saussure's components here to define what I later call a polyvalent symbol rather than a sign. But, departing from Saussure, I will show that the sound images I focus on may actually point to several different concepts.

2. Here I fuse the semiologist's notion of the sign with the field of semantics to create an understanding of the multiple conceptual fields that a single acoustic image can point to. Saussure never dealt with the subject of semantics, so I draw on more recent work in ethnobotany, ethnobiology, and ethnoecology, particularly the work of Conklin (1954), Kay (1971), Ellen and Reason (1979), and Berlin (1992). These scholars have developed an understanding of the semantic fields and taxonomic properties of botanical and biological classifications articulated by diverse cultural groups. While I am not concerned here with taxonomic principles, their accounts are instructive for the detailed understanding of semantic domains they provide.

3. See also Ellen's (1982:206–26) critique of the analysis of classifications in the study of cognitive models of the environment. He points out that, while important to the study of perceptions of the environment, the construction of classificatory systems should also include attention to the social context of particular classifications and to the techniques and statements of knowledge and practice. Moreover, Ellen argues that we should be concerned with the way in which the cognitive organization of recognized environmental phenomena affects ecological interactions (1982:210) and mentions limitations of this model: (1) that individual statements cannot be immediately abstracted to represent an entire culture or the practical applications that individuals have for their ideal classifications, and (2) that a static model of classification cannot portray the fluidity of category boundaries or the recognition of continua, succession, tendencies, and cycles (see also chap. 1). I would add here that static categories do not suggest perceptions of transformation unless a new kind of categorization based explicitly on these perceptions is elicited and charted.

4. Here Bourdieu's (1977) notion of disenchantment is used to refer to what van der Veer (1988) and Parry (1994) describe as the this-worldly nature of the otherworldly goals of pilgrim service and service to deities.

5. On one occasion an American scholar on my behalf posted a question on the internet about the meaning of *paṇḍāgīrī* to a listserve of religious studies scholars and experts. Many responded that the term literally meant "paṇḍā's work" but then made more figurative references to its meaning as shady people doing shady deals.

6. The worship rituals of pūjā, śrāddha, and tarpaṇ are done with sacred Gaṅgā water. Pūjā rituals are done to please particular gods and goddesses, while śrāddha and tarpaṇ are done to give respect to the ancestors. Many have written on the meaning and specific elements of these rituals (see, e.g., Eck 1981; Fuller 1979; Parry 1994).

7. Dān in this context is not as debilitating as Parry (1980:102) suggests it once was for the Mahābrahmans of Maṇikarṇikā. If it ever was, dān today is not feared for the "pollution" it transfers from giver to receiver.

8. Actually, rights to serve pilgrims are highly regulated on some ghāṭs. This regulation is meant to ward off free-market competition in pilgrim services. Parry (1994) presents a detailed discussion of *pārī*, the hereditary-market system of rights followed to organize services on Maṇikarṇikā ghāṭ. Pārī rotates service responsibilities through many claimants.

9. I am arguing here that informants include uncremated or partially cremated bodies in the semantic domain of *gandagī* because of changes in ritual practices and exposure to meanings attributed to these corpses by media and official agents. However, I place these bodies at the margins of that semantic category, since they are not what Berlin (1992:25) has called prototypes. A prototype is a focal member of a category and is seen as the best example of that category. Instead, corpses lie at the boundaries of this domain, claiming some affinity with elements inside the domain and some proximity to elements in other domains, especially to those in the domain of impurities purified through ritual practices (a domain I do not elaborate in graphic form here).

10. *Amrit* is the Hindi term for sacred nectar, a nectar often deposited by gods and goddesses in the earth to symbolize their contact with this world. This nectar brings immortality, release from the bondage of rebirth, and nourishment for the soul.

11. These statements were drawn from a tape-recorded interview with this individual.

CHAPTER 5

1. His words were: *koi praduṣan samajh mai nahī ātā yahā par.*

2. See Hunter 1888:1–5.

3. As I explained in chapter 4, a *ghāṭ* is a flight of steps that provides access to the river. It is also considered a place of sacred power.

4. This is the site of the *kuṃbha melā*, a festival that takes place once every twelve years in Haridwar, Allahabad, Nasik, and Ujjain.

5. See Hamilton 1993:667–69.

6. Hunter 1888:1–5; Varun 1981:336.

7. Over the last quarter of the nineteenth century, irrigation canals in Roorkee district *(tahsil)* effectively combated drought (see Varun 1981:112). In contrast to 1868–69, when almost all the autumn crops were lost except those in the irrigated region, in 1870 162,317 acres in Saharanpur district were irrigated out of which 84,404 (more than half) were watered from canals alone (see Hunter 1888:104). In 1884–85, canals supplied water to 71,916 acres (29,100 hectares). This increased to 80,724 acres (32,668 hectares) in the next decade and to 121,550 acres (49,190 hectares) in the following year.

CHAPTER 6

1. For exceptions, see Derman 1998; Miller et al. 1996.

2. Environmental racism, introduced in the famous study conducted by the United Church of Christ in 1987, points attention to the ways in which communities of racial minorities are exposed to a disproportionate amount of environmental hazards. Studies by Robert Bullard (1983, 1992, 1993) brought sociology closer to this field of study and eventually led to policy changes at the Environmental Protection Agency and to the creation of their environmental justice program (see also Bailey and Faupel 1992a, 1992b; Bailey et al. 1993; Bryant and Mohai 1992; United Church of Christ 1987; Miller 1993; Novotny 2000; Szasz 1994). One of the few studies that looks at legal defenses organized by grassroots groups concerned with environmental racism is Miller et al.'s (1996) study of resistance to a sewage treatment plant in Harlem, New York.

3. British colonial administrators, like their counterparts in England, saw forests as resource-rich frontiers with great market potential. The image of the frontier is evident in the Indian Forest Act of 1865, which set out the state's rights to forest tracts. In the amendments that followed, the Forest Act brought more and more forest acreage under the purview and control of government as it tried, sometimes with great struggle, to limit villagers' and forest dwellers' uses of its products (see Baviskar 1995; Guha 1983, 1994; Gadgil and Guha 1992; Rangan 1996).

4. In a case tried in 1884, the judicial authorities defined "public" as follows: "A place is a public place if people are allowed access to it, though there may be no legal right to it. So, a well is a public well if people are allowed to use its water" (*Reg. v. Wallard* 1884, in *All India Reporter* 1916:15).

5. The cases cited are 2 BomLR 107 (1900:1079)** and 13 CPLR 92 (1900:93).

6. *Bombay Law Review* (1900:1078). Criminal reference in *Queen-Empress v. Bhagi kom Nathiaba* Indian Penal Act XLV of 1860 Sec. 277. The language in this citation is rather difficult to understand, especially the references to the taking of water by higher castes. The citation seems to say that higher castes permitted other higher castes but would not permit her, as a member of a lower caste, to take the water. If she were to take it, then the higher-caste members would no longer be able to drink from the well without risk of pollution.

7. *Bombay Law Review* (1900:1078).

8. This was in line with their precept of "equality before the law." The status and ties of parties and matters of importance to an indigenous council were deliberately ignored (Galanter 1989:19).

9. Cited in *All India Reporter* 1924:121.

10. *Kutti Chami Moothan v. Rama Pattar*, in *All India Reporter* 1919:755.

11. *Kutti Chami Moothan v. Rama Pattar*, in *All India Reporter* 1919:755.

12. *All India Reporter* 1924:121.

13. *Emperor v. Nama Rama* in Chaudhri 1904:6; *The Empress v. Halodhur Poroe and others* in *Indian Law Reports* 1876:383. Other more cryptic references include: (1904) I CriLJ6 (Bom), (1898) 3 MysCR No. 172. p.509 (510) (DB) xx(1882) 1LR 4 Mad 229 (229) xx (1876–1877) ILR 2 Cal 383 (383) xx 1869 Rat Un Cr Cas 14(14).

14. *Air Man*, vol. 28: 255–60, on Section 268.

15. The main aim of the act was to prohibit any alteration in the flow of water in a river that might cause damage to a canal or drainage work.

16. The Factories Act of 1948 directed factories to make arrangements for the treatment of wastes and effluents according to rules framed by the state government (Act LXIII of 1948; Manohar and Chitaley 1979c:781; Mathur 1980:92). Finally, the river boards established under the River Boards Act of 1956 to regulate the flow of rivers for irrigation and water supply were empowered to prevent pollution (Act 49 of 1956; Manohar and Chitaley 1979d:407; Rosencranz et al. 1991:38).

17. This reference is taken from the Supreme Court case *M. C. Mehta v. Union of India*, AIR SC 1086 (1987), better known as the Delhi Gas Leak Case. This particular quotation is reproduced from Mehta n.d.:11. The rule of Absolute Liability that Justice Bhagwati developed specifically targeted hazardous and inherently dangerous industries that threatened the health and safety of persons working in the factory or residing in the surrounding area. The rule stipulated that the industry is responsible for absorbing the cost of any accident arising from that hazardous or dangerous activity.

CHAPTER 7

1. The Environment (Protection) Act, 1986 No.29 of 1986; cited in Central Pollution Control Board 1995c:214.

2. Mandamus is a judicial remedy in the form of an order from a superior court to any government, court, corporation, or public authority. The order requires the mentioned party to do some specific act which that body is obliged under law to do or refrain from doing what is not allowed.

3. I collected data on these cases during several stays in Kanpur and New Delhi from 1994 through 2001. Some Supreme Court orders I cite were taken from legislative digests, while others were culled from the files of environmental lawyers and government departments. Information provided in these orders is supplemented by my own descriptions of the agencies working under the Ganga Project Directorate in the Ministry of Environment and Forests. Discussions with environmental activists, foreign consultants, and local residents also helped me

outline the wastescapes that are referred to in the legal debate and that are targeted by institutional projects.

4. See Rosencranz et al. 1991:184; Singh et al. n.d.:194–95. The Indo-Dutch Environmental and Sanitary Engineering Project (hereafter Indo-Dutch Project) was conceived under the first phase of the Ganga Action Plan to administer the works undertaken with a grant provided by the Dutch government to the government of India. The works under this name were considered part of the Ganga Action Plan for Kanpur and Mirzapur. Although this paper does not discuss projects in Mirzapur, the Indo-Dutch projects in both these cities were linked. The total investment provided by the Dutch government was NLG 27.9 million or 356 million rupees in Kanpur and 202 million rupees in Mirzapur.

5. Personal communication with Mr. M. C. Mehta.

6. Many of the individual Court orders described here are not listed in law journals, nor are they cited yet in the volumes of the Supreme Court cases. I obtained copies of these orders from the files of M. C. Mehta. This particular order was made for the writ petition (C) No. 3727 of 1985 on 7 October 1994.

7. Supreme Court Order made from the writ petition (C) No. 3727 of 1985, 20 July 1995, 8.

8. Supreme Court Order made from the writ petition (C) No. 3727 of 1985, 20 October 1995, 5–6.

9. World Health Organization 1984–85.

CHAPTER 8

1. Duncan to Government, 22 May 1791, Consultation 3 June 1791, Bengal Revenue Judicial Range 127, vol. 71, 29–32. See Arnold 1989:264 for these citations.

2. Duncan to Government, 17 October 1792, Consultation 29 October 1792, Bengal Revenue Judicial Range 127, vol. 82, 493–94.

3. British measures to deal with waste were part of a larger imperial vision that manifested itself in different ways in other colonial projects. Argryous (1997:168) explains that the measures the British took to eradicate litter in Cyprus were part of a wider scheme to extricate Cyprus from the web of its Ottoman past.

4. Lok Sabha Debates, 7 August 1985.

5. Lok Sabha Debates, 20 November 1985, 4 December 1985, 30 July 1986, 6 August 1986, 25 February 1987, 3 December 1986.

6. Lok Sabha Debates, 25 February 1988.

7. The power supply in Uttar Pradesh is very erratic, and in Banaras the supply is usually broken by several outages a day. While statistics are not available, in any twenty-four-hour period, power is not available for perhaps five to six hours. This blackout period increases during the hot summer months and the coldest winter month.

8. Moreover, even under optimum working conditions, the activated sludge process cannot reduce 90 percent of the bacterial load in the treated sewage (Ministry of Environment and Forests 1998:19). This means that fecal coliform counts may be as high in the treated effluent as they are in the influent.

9. See Gupta 1995 and Monteiro 1966 for a discussion of corruption in Indian bureaucracy and Oldenburg 1987:328–29 for a more specific reference to corruption in construction projects. Construction projects tend to be a favorable avenue for what he calls systematic corruption.

10. BMB et al. 1995b:8. See also Bhattacharya 1974, Hussain 1989, and Srivastava 1980 for a discussion of the historical problems of municipal administration in India.

11. See Potter 1996 for a discussion of the evolution of this policy of transference in the Indian Civil Service and later in the Indian Administrative Service.

12. Alongside the Dutch project in Kanpur, the government of India entered into an agreement with the World Bank to provide funding for a 130 mld sewage treatment plant in the same city. The fact that two different projects funded by two different sources exist side by side may affect the extent to which the Indo-Dutch project can succeed in its institutional strengthening objectives.

13. On the other hand, the Jal Nigam attempts to back out of the less enviable tasks of *solid waste* management because they have not yet come under the purview of foreign-funded projects. Solid waste management is a field that has thus far attracted little in the way of grants and loans because it does not involve heavy construction. Solid waste is left to accumulate and disintegrate on vacant land, while wastewater is managed by expensive disposal and treatment systems. But as vacant land becomes more scarce, we may find the politics of land acquisition assuming a bigger role in solid waste management projects and raising the bureaucratic desire for managing solid waste sites. Moreover, recent work by the Central Pollution Control Board to formulate a policy on hazardous waste management indicates that foreign interests may soon turn in that direction, to provide technical assistance and new liner technologies.

14. Among the organizations making these allegations are the National Environmental Engineering Research Institute in Kanpur and the Regional Centre for Extension and Development of the Central Leather Research Institute of Kanpur.

15. It is also possible that industries were the first to set this precedent for behavior, especially for the practice of paying off government officials to avoid compliance to regulations and procedures. Whoever started it, the practice now locks together the interests of industrialists and government regulatory officials and creates a nexus so strong that the most stringent court orders cannot break it. In fact, as this case shows, court orders may strengthen that nexus by giving it the impetus for collecting fees that allows noncompliance.

CHAPTER 9

1. *Times of India*, 19 September 1992.

2. See *Indian Express*, 21 January 1992, and *Hindustan Times*, 12 July 1992, for reports on monitoring exercises carried out by the SMF, the Jal Nigam, the Uttar Pradesh Water Pollution Control Board, and the Central Water Commission. These reports revealed BOD levels over 5.5 and fecal coliform counts of 5,000 to 10,000 per 100 milliliters. The *Hindu*, 13 April 1992, cited results of SMF studies of total coliform count of 24,000 MPN per 100 milliliter and a fecal coliform

count of 12,000 MPN per 100 milliliters. BOD levels cited in other reports varied from 8.7 milligrams per liter to 12 milligrams per liter at various ghāṭs and was as high as 505 milligrams per liter downstream from Khirkī nālā (see also Tiwari and Sen 1991:78). Maximum permissible numbers of fecal coliform cells for bathing are set by the government at 500 cells per 100 milliliters. Acceptable BOD levels for bathing range from 1 to 2 milligrams per liter. For drinking, both parameters ought to be significantly lower.

3. *Hindu*, 15 January 1992; *Hindustan Times*, 15 January 1992.

4. *Times of India*, 10 April 1999.

5. The Eco-Friends report was handwritten and contained maps to mark where the data were observed and collected. I verified their claims through field surveys. Because this voluntary organization was not equipped to measure effluent discharge, most of the claims were made on the basis of visible evidence such as the signs of the flow of sewage and tannery wastewater. The intention of their report was to register a layperson's view of sources of pollution in the Ganga.

6. Two electric crematoriums were built in the Kanpur area under the first phase of the Ganga Action Plan to provide an alternative to the high cost of wood cremation. However, poor acceptance by the public coupled with opposition by owners of the wood crematoriums has resulted in low usage levels. In the three years following completion in 1991, one crematorium cremated 313 bodies and the other 74 (Ministry of Environment and Forests 1995:19). Both are now maintained by the Kanpur Municipal Corporation. A third electric crematorium, constructed before GAP began, has been closed for the past seven years due to a break in the platform construction.

7. Justice Bhagwati, former chief justice of India, articulated these concerns when he said, "The Court will not insist on a regular writ petition and even a letter addressed by a public spirited individual or a social action group acting 'probono publico' would suffice to ignite the jurisdiction of this court" (Mehta n.d.:6–9). Other environmental cases in which liberal precedents for locus standi and epistolary jurisdiction have been established are *S.P. Gupta v. Union of India*, Supp. SCC 87 (1981) and *M.C. Mehta v. Union of India*, AIR SC 1086 (1987) (see also Jariwala 2000). Through personal communication with this individual, I learned of his intention to parallel the strategies taken in the Ganga Pollution Cases.

8. This estimate was extrapolated from my field visits at various times of the day. I also have numerous pictures of the flow from Śivālā nālā, taken at various times over the past five years.

9. I also found that by February 2000, Śivālā nālā had grown significantly and now has a hefty perennial flow.

10. For instance, in a Supreme Court order dated 8 July 1996, 168 factories categorized as 'H' (noxious and hazardous) were relocated out of the nation's capital of New Delhi and 50,000 workers lost their jobs in those plants (*M.C. Mehta v. Union of India* No. 4677 of 1985). Many reports and surveys conducted during and after these relocations found that industries had flagrantly violated the worker compensation provisions of the Supreme Court orders and simply left many laborers without adequate alternative employment. Those orders also favored industrialists and real estate agents who were able to take over the aban-

doned plots at low prices and develop them into fashionable spaces for the urban elite (see A Correspondent. "Living on the Edge: Human Cost of Closures of Industries," *Economic and Political Weekly*, 11 April 1998:817). These orders emboldened, as NGOs have put it, the government and the land mafia. Though the court orders relocated polluting industries, they did not guarantee a reform in their future behavior.

CHAPTER 10

1. See Chapple and Tucker 2000; India International Centre 1992; Mumme 1998; Nagarajan 1998; Narayanan 2001; Nelson 1998; Sherma 1998.

2. These statements, provided by a svāmī of Rama Krishna Mission, were tape-recorded during an interview at the melā in January 1996.

3. From tape-recorded interview with Śaṅkarācārya Shri Nischalanand of Saraswati Goverdhan Muth in Jagannath Puri.

4. See Marglin's (1977:252) discussion of Lingat 1973.

5. Excerpt from a tape recording of his talk. The talk was given in Hindi, but the phrases in *italic* mark where he spoke in English.

6. The memorandum later stated that the aim was "to get people [to] take pledge to construct stately temples symbolizing our national heritage and moorings at Sri Ram Janma Bhumi, Ayodhya, Sri Krishna Janmasthan, Mathura, and Sri Gyaanvaapi Vishwanath spot, Varanasi" (Vishwa Hindu Parisad 1995:2).

7. Present at the occasion were Madhya Pradesh Minister of Parliament Swami Chinmayananda, Ashok Singhal, Vishnu Hari Dalmia, Sankaracarya Rambhadracarya of Chitrakoot, Sankaracarya Jagadguru Ramanandacarya Ji (the identity of both as Śaṅkarācāryas is disputed), Haryacarje Maharaj of Ayodhya, Swami Prakashanand Giri Mahasnandedeshwar, and Jagadguru Vasudevanand Saraswati.

8. Burakoti 1996:24. *Pajcjanya*, the name of the VHP newsletter that published this article, is the name of Kṛṣṇa's conch.

9. Other participants were Sri Ashok Singhal, president of VHP, Rajmata Vijayaraje Scindia, member of Parliament from Madhya Pradesh, Susri Uma Bharti, member of Parliament from Madhya Pradesh, Dr. Ram Bilas Vedanti, member of Parliament from Uttar Pradesh, and Mahant Avaidya Nath, member of Parliament from Uttar Pradesh. These individuals constitute most of the members of Parliament who were religious leaders and VHP activists at the time. Other religious leaders participating in the yatra were Swami Vasudevanand Saraswati, Śaṅkarācārya of Badrik Ashram in Badrinath, and Susri Shiva Saraswati, a firebrand VHP activist from Rajasthan. All the MPs and MLAs belonging to the BJP in each of the cities they visited en route joined the yatra in their respective locales.

10. Swami Chinmayananda's statements were given to me by Eco-Friends members who accompanied the VHP on the entire journey. These statements come from an undated and unpublished report written by Swami Chinmayananda.

CHAPTER 11

1. Consider the debates and riots in December 2000 on Delhi streets as the Supreme Court attempted to enforce a 1995 order to relocate small and large industrial units located in the nonconforming areas of the city.

2. This sensitivity is also raised in recent debates about environmentalism (see Brosius 1999a). Essentializing practices have been a part of official and grassroots environmental movements for some time, some of them embedded in romantic images of community and nature and others employed as strategies for public consciousness-raising and mobilization.

3. This contrasts with Feldhaus's (1995) work on rivers in Maharastra. Interviewing many women, Feldhaus found that they tended to stress the fertility power of rivers over their purificatory power, a theme that may reflect their own knowledge of fertility and their own associations of fertility with power.

4. This characteristic can be contrasted with those of other gods and goddesses who become angry if defiled by humans. See Gold 1998; Sherma 1998; Nagarajan 1998; Dumont and Pocock 1959:31; Harper 1964:183–86; Srinivas 1952:41–42, 78.

5. This tension parallels the tension that existed in seventeenth- and eighteenth-century European philosophy. In the European tradition, this was a tension between discourses that took nature as a theological given and philosophies that argued for the primacy of natural law and an opposition between reason and nature (Lease 1995:8–9). The earlier European position on nature as a theological given and the contemporary Hindu position on sacred ecology are, however, different in one important respect. European Christian philosophers made nature an act of divine creation and defined that theological given in contradistinction to the god of creation.

6. Two of the public interest cases outlined here clearly rose and fell on the charisma of key justices, providing evidence that judicial activism has limits to its effectiveness and efficacy as a tool against the state. This makes the work of the petitioner and the nongovernmental organization all the more challenging, since achievements made through legal orders and judgments can be manipulated by outside committees and reviewers or reversed by simple inaction and noncompliance. While both the cases I summarized have yet to be concluded, they continue to weather periods when the judiciary takes little interest. During these periods, petitioners and lawyers turn to work at other levels, participate in other regulatory and oversight committees, and organize public awareness programs and street activism. This is the kind of versatility required of the public interest lawyers and NGOs I have introduced.

7. Accounts of public interest litigation should be included in the nascent field of public interest anthropology. Yet this diffuse "public interest" that anthropologists and lawyers refer to has to be carefully defined and the specific interests of public coalitions qualified in their appropriate contexts. The outward layering method I have employed to get at intersecting discourses is one way in which the diversity of public interests can be analyzed. The reason that public interest litigation is important to the field of environmental activism is that it deals with confrontations to domination, while using the written rules of the nation-state. I have tried to make this dynamic more specific by showing that this particular public interest lies in the intentions of petitioners and activist judges and in the coalitions that form in and through these legal gateways.

8. Upendra Baxi has written many articles about judicial activism (see Baxi 1998a, 1998b, 1998c, 2000). Preferring the term social action litigation (SAL) over the term public interest litigation (PIL), Baxi distinguishes the American

experience with PIL from the emergence of judicial activism in India (see 1998a:A-91–A-93). He argues that the American PIL movement "involved innovative uses of the law, lawyers and courts to secure greater fidelity to the parlous notions of legal liberalism and interest group pluralism in an advanced industrial capitalistic society" (1998a:A-92). Indian SAL, on the one hand, is judge-led and judge-induced and, on the other, inspired by citizens' bold petitions to, as Baxi (1998a:A-97) put it, take human suffering seriously. I have used PIL instead of SAL in this text because the term PIL is more widely used in the Indian judicial literature, the media, and discussions among the activists I have covered. But Baxi's call to distinguish Indian from American public interest litigation is extremely valuable and needs to be taken up in future research.

9. See also Martinez-Alier 1998. While ecological debts have been discussed largely in terms of the damages to ecosystems caused by consumers and producers in Western countries (Agarwal and Narain 1991; Azar 1994), the notion can also be used to describe any contribution to the waste stream that originates from a local source and then spreads outward. With reference to air, Bellows (1996) talks about the ecological debt incurred in the sulfur triangle from cities that span the four countries of Poland, the Czech and Slovak Republics, and (the former East) Germany.

Glossary

akhāṛā	wrestling ground
āratī	oil lamp ritual
bhaktī	devotion to the divine
dāna	donation
darśan	auspicious sight of the divine
gand, gandagī	waste or matter out of place
ghāṭ	sacred space adjoining river; the steps leading down the bank
ghāṭiyā	lower ranked pilgrim priest
jala	sacred water
jāti	caste
Kali yuga	fourth eon of time spelling total decline of morality and humanity
kuṅḍ	tank of water
liṅga	the iconic, phallic form of Siva
mantra	sacred phrase or chant
melā	religious festival
mokṣa	final liberation
nālā	wastewater drain
nāstik	unbeliever
pakkā	neat and accurate
pān	betelnut chew
paṇḍā	pilgrim priest
peśā	occupation
praduṣan	pollution (Hindi transliteration of 'pollution')
pūjā	worship ritual
purohit	priest
rishī	monk, renouncer
sādhu-sant	world renouncer and saint
saṃsāra	cycles of reincarnation
Śaṅkarācārya	respected Hindu religious leader
śāstra	sacred text

snān	ritual ablution
śrāddha	ritual for the ancestors
stuti	hymn
śuddha	pure
svacchtā	cleanness
tarpaṇ	libation to the ancestors
tīrth purohit	pilgrim priest
tīrtha	sacred place
yātra	pilgrimage

Bibliography

SACRED TEXTS, WITH ABBREVIATIONS

BhP *Bhāgavata Purāṇa.* Translated and annotated by G. V. Tagare. Part II (Skandas 4–6). Delhi: Motilal Banarsidass, 1976.

KKh *Kāśī Khaṇḍa.* Ed. A. S. K. Tripathi. 2 parts. Varanasi: Sampurnanand Sanskrit University, 1991. All translations cited are from Singh 1993a.

PM *Paramśānti Kā Mārg.* Shri Jaidayaal Goyandka. Varanasi. All translations are my own.

SP *Śiva Purāṇa.* Translated by a board of scholars; ed. J. L. Shastri. Vol. I. Delhi: Motilal Banarsidass, 1990.

REFERENCES

Abu-Lughod, Lila

1986 *Veiled Sentiments: Honor and Poetry in a Bedouin Society.* Berkeley: University of California Press.

Adas, M.

1992 From Avoidance to Confrontation: Peasant Protest in Precolonial and Colonial Southeast Asia. In *Colonialism and Culture*, ed. Nicholas Dirks, 89–126. Ann Arbor: University of Michigan Press.

Adler, Robert

1999 Integrated Approaches to Water Pollution: Lessons from the Clean Air Act. *Harvard Environmental Law Review* 203:203–95.

Agarwal, Anil

1994 An Indian Environmentalist's Credo. In *Social Ecology*, ed. Ramachandra Guha, 346–84. Delhi: Oxford University Press.

2000 Can Hindu Beliefs and Values Help India Meet Its Ecological Crisis? In *Hinduism and Ecology*, ed. Christopher Key Chapple and Mary Evelyn Tucker. Cambridge: Harvard University Press.

Agarwal, Anil, and Sunita Narain

1991 *Global Warming in an Unequal World: A Case of Environmental Colonialism.* Delhi: Centre for Science and Environment.

Alcoff, Linda

1994 Cultural Feminism versus Post-Structuralism: The Identity Crisis. In *Culture/Power/History: A Reader in Contemporary Social Theory*, ed. Nicholas Dirks, G. Eley, and Sherry Ortner, 96–122. Princeton: Princeton University Press. (Reprinted from *Signs: Journal of Women in Culture and Society* 13, no. 3:405–36.)

All India Reporter

1916 Nagpur Section. Nagpur: V. V. Chitaley.

1919 Nagpur Section. Nagpur: V. V. Chitaley.

1924 Nagpur Section. Nagpur: V. V. Chitaley.

1988 Vol. 75. Supreme Court Section. Nagpur: V. V. Chitaley.

Alley, Kelly D.

1990 *Tourism in India: Marketing Culture*. Ph.D. thesis, University of Wisconsin-Madison.

1992 On the Banks of the Ganga. *Annals of Tourism Research* 19, no. 1:125–27.

1994 Ganga and Gandagi: Interpretations of Pollution and Waste in Benaras. *Ethnology* 33, no. 2:127–45.

1996 Urban Institutions at the Crossroads: Judicial Activism and Pollution Prevention in Kanpur. *Urban Anthropology* 25, no. 4:351–84.

1997 Gandhiji on the Central Vista: A Postcolonial Refiguring. *Modern Asian Studies* 31, no. 4:967–94.

1998 Idioms of Degeneracy: Assessing Ganga's Purity and Pollution. In *Purifying the Earthly Body of God: Religion and Ecology in Hindu India*, ed. Lance E. Nelson, 297–330. Albany: State University of New York Press.

2000 Separate Domains: Hinduism, Politics and Environmental Pollution. In *Hinduism and Ecology*, ed. Christopher Key Chapple and Mary Evelyn Tucker. Cambridge: Harvard University Press.

Alley, Kelly D., Conner Bailey, and Charles Faupel

1995 The Historical Transformation of a Grassroots Environmental Group. *Human Organization* 54, no. 4:410–16.

Alter, Joseph

1992 *The Wrestler's Body: Identity and Ideology in North India*. Berkeley: University of California Press.

Apffel-Marglin, Frederique

1992 Women's Blood: Challenging the Discourse of Development. *Ecologist* 22, no. 1:22–32.

1994 *Decolonizing Knowledge: From Development to Dialogue*. Oxford: Clarendon.

Apffel-Marglin, Frederique, and Stephen Marglin, eds.

1990 *Dominating Knowledge: Development, Culture and Resistance*. Oxford: Clarendon.

Appadurai, Arjun

1988 Putting Hierarchy in Its Place. *Cultural Anthropology* 3, no. 1:36–49.

1990 Disjuncture and Difference in the Global Cultural Economy. *Public Culture* 2, no. 2:1–24.

1991 Global Ethnoscapes: Notes and Queries for a Transnational Anthropology. In *Recapturing Anthropology: Working in the Present*, ed. Richard Fox, 191–210. Santa Fe: School of American Research.

Argryous, Vassos

1997 Keep Cyprus Clean: Littering, Pollution and Otherness. *Cultural Anthropology* 12, no. 2:159–78.

Arnold, David

1989 The Ecology and Cosmology of Disease in the Banaras Region. In *Culture and Power in Banaras*, ed. Sandria Freitag, 246–67. Berkeley: University of California Press.

Arnold, David, and Ramachandra Guha, eds.

1995 *Nature, Culture, Imperialism: Essays on the Environmental History of South Asia*. Delhi: Oxford University Press.

Audit Report on Ganga Action Plan in Uttar Pradesh: Gomukh to Kannauj. New Delhi: Akashdeep Printers.

Azar, Christian, and J. Holmberg

1994 Defining the Generational Environment. *Ecological Economics* 14:7–20.

Babb, Lawrence

1975 *The Divine Hierarchy: Popular Hinduism in Central India*. New York: Columbia University Press.

Bailey, Conner, and Charles E. Faupel

1992a Environmentalism and Civil Rights in Sumter County, Alabama. In *Race and the Incidence of Environmental Hazards*, ed. Bunyan Bryant and Paul Mohai, 140–52. Boulder: Westview.

1992b Movers and Shakers and PCB Takers: Hazardous Waste and Community Power. *Sociological Spectrum* 13:89–115.

Bailey, Conner, Charles E. Faupel, and James Gundlach

1993 Environmental Politics in Alabama's Blackbelt. In *Confronting Environmental Racism: Voices from the Grassroots*. Ed. Robert Bullard, 107–22. Boston: South End Press.

Bailey, Conner, Charles E. Faupel, Susan Holland, and Amy Waren

1989 *Public Opinions and Attitudes Regarding Hazardous Waste in Alabama: Results from Three 1988 Surveys*. Auburn: Alabama Agricultural Experiment Station.

Bakuniak, Grezegorzi, and Krzysztof Nowak

1987 The Creation of a Collective Identity in a Social Movement. *Theory and Society* 16:401–29.

Balasubramaniam, Vejai

1998 Environment and Human Rights: A New Form of Imperialism? *Political and Economic Weekly*, 21 February, 389–90.

Balee, William

1998a *Advances in Historical Ecology*. New York: Columbia University Press.

1998b Introduction to *Advances in Historical Ecology.* New York: Columbia University Press.

Bandyopadhyay, J., N. D. Jayal, U. Schoettli, and Chhatrapati Singh, eds.

1985 *India's Environment: Crisis and Responses.* Dehra Dun: Natraj Publishers.

Basu, Kaushik

1999 Budget of Enterprise. *India Today International* (New Delhi), 1 March, 31.

Baviskar, Amita

1995 *In the Belly of the River: Tribal Conflicts over Development in the Narmada Valley.* Delhi: Oxford University Press.

1999 Comment on "Analyses and Interventions: Anthropological Engagements with Environmentalism," by J. Peter Brosius. *Current Anthropology* 40, no. 3 (June):277–309.

Baxi, Upendra

1985 *Courage, Craft and Contention: The Indian Supreme Court in the Eighties.* Bombay: Tripathi.

1998a Taking Suffering Seriously: Social Action Litigation in the Supreme Court of India. In *Supreme Court on Public Interest Litigation,* ed. Jagga Kapur, vol. 1, A-91–A-114. Delhi: Lips Publication.

1998b Judicial Activism: Usurpation or Re-democratization? In *Supreme Court on Public Interest Litigation,* ed. Jagga Kapur, vol. 1, A-131–A-144. Delhi: Lips Publication.

1998c Towards a Structural Adjustment of Judicial Activism? In *Supreme Court on Public Interest Litigation,* ed. Jagga Kapur, vol. 1, A-145-A–A-145-X. Delhi: Lips Publication.

2000 The Avatars of Indian Judicial Activism: Explorations in the Geographies of [In]Justice. In *Fifty Years of the Supreme Court of India,* ed. S. K. Verma and Kusum, 156–209. Delhi: Oxford University Press.

Beder, Sharon

1989a The Effects of Environmental Regulation on the Development of Waste Management Technology. *Planning for Environmental Change* no. 89/3:124–28.

1989b Professional Influence and Public Effluent. *Australian Society,* May, 71–72.

1989c Engineering Sydney's Sewerage Pollution: Public Relations Assisted Technology. *Current Affairs Bulletin* 66, no. 2:27–31.

1990a The Sydney Experience. In *Sydney's Strangled Sewerage System,* ed. A. M. Cooke. The Australian and New Zealand Association for the Advancement of Science.

1990b Early Environmentalists and the Battle against Sewers in Sydney. *Royal Australian Historical Society Journal* 76, no. 1:27–44.

1990c Surf and Seafood Come under Suspicion in Sydney. *New Scientist* 14:24–29.

1990d Science and the Control of Information: An Australian Case Study. *Ecologist* 20:136–40.

1991a Recycling Innovations in Sewage Treatment. *Search* 21, no. 8:247–48.

1991b Controversy and Closure: Sydney's Beaches in Crisis. *Social Studies of Science* 21:223–56.

1992 Getting into Deep Water: Sydney's Extended Ocean Sewage Outfalls. In *A Herd of White Elephants: Australia's Science and Technology Policy*, ed. Pam Scott, 62–74. Sydney: Hale and Iremonger.

1993a Pipelines and Paradigms: The Development of Sewerage Engineering. *Australian Civil Engineering Transactions* CE35, no. 1:79–85.

1993b From Sewage Farms to Septic Tanks: Trials and Tribulations in Sydney. *Journal of the Royal Australian Historical Society* 79, nos. 1 and 2:72–95.

1995 Sewage Treatment and the Engineering Establishment. In *Confronting the Experts*, ed. Brian Martin, 13–44. Albany: State University of New York Press.

Bellows, Anne C.

1996 Where Kitchen and Laboratory Meet: The "Tested Food for Silesia" Program. In *Feminist Political Ecology: Global Issues and Local Experiencs*, ed. Dianne Rocheleau et al., 251–70. New York: Routledge.

Berkes, F.

1985a The Common Property Resources Problem and the Creation of Limited Property Rights. *Human Ecology* 13, no. 2:187–208.

1985b Fishermen and "The Tragedy of the Commons." *Environmental Conservation* 12, no. 3:199–209.

Berlin, Brent

1992 *Ethnobiological Classification: Principles of Categorization of Plants and Animals in Traditional Societies.* Princeton: Princeton University Press.

Bhargava, D. S.

1981a Ganga, the Most Self-Purifying River. In *International Symposium on Water Resources Conservation, Pollution and Abatement.* December 11–13. Department of Civil Engineering, University of Roorkee. Delhi: Sarita Prakashan.

1981b Pollution Control Strategy for Ganga. In *International Symposium on Water Resources Conservation, Pollution and Abatement.* December 11–13. Department of Civil Engineering, University of Roorkee. Delhi: Sarita Prakashan.

Bhattacharya, Mohit

1974 *Municipal Government: Problems and Prospects.* Delhi: P. Jain.

Blair, Harry

1996 Democracy, Equity and Common Property Resource Management in the Indian Subcontinent. *Development and Change* 27:475–99.

Blumer, Herbert

1939 Collective Behavior. In *Principles of Sociology*, ed. Alfred Lee. New York: Barnes and Noble.

BMB Management, JPS Associates, and Euroconsult
1995a *Institutional and Community Development Project, Phase One: Technical Report* no. 8 (June).
1995b *Ganga Institutional and Community Development Project: Inception Report.*
Borrelli, Peter
1987 Environmentalism at a Crossroads. *Amicus Journal* 9, no. 3:24–A37.
Bourdieu, Pierre
1977 *Outline of a Theory of Practice.* Cambridge: Cambridge University Press.
Brass, Paul
1997 *Theft of an Idol.* Princeton: Princeton University Press.
Breckenridge, Carol, and Peter van der Veer
1993 *Orientalism and the Postcolonial Predicament.* Philadelphia: University of Pennsylvania Press.
Bromley, Daniel
1992 *Making the Commons Work.* San Francisco: Institute for Contemporary Studies.
Brosius, J. Peter
1999a Analyses and Interventions: Anthropological Engagements with Environmentalism. *Current Anthropology* 40, no. 3 (June): 277–309.
1999b Green Dots, Pink Hearts: Displacing Politics from the Malaysian Rain Forest. *American Anthropologist* 101, no. 1:36–57.
Bru-Bistuer, Josepa
1996 Spanish Women against Industrial Waste. In *Feminist Political Ecology: Global Issues and Local Experiences,* ed. Dianne Rocheleau, Barbara Thomas-Slayter, and Esther Wangari, 105–24. New York: Routledge.
Bryant, Bunyan, and Paul Mohai, eds.
1992 *Race and the Incidence of Environmental Hazards.* Boulder: Westview.
Bryant, Raymond L., and Sinead Bailey
1997 *Third World Political Ecology.* London: Routledge.
Bullard, Robert D.
1983 Solid Waste Sites and the Black Houston Community. *Sociological Inquiry* 53, nos. 2–3:273–88.
1992 *Dumping in Dixie: Race, Class and Environmental Quality.* Boulder: Westview.
1993 *Confronting Environmental Racism: Voices from the Grassroots.* Boston: South End Press.
Burakoti, Padramesh
1996 The Construction of the Tehri Dam Continues at the Cost of Ruining Our Condition. *Pajcjanya,* 24 September, 24.
Burke, Kenneth
1955 *A Grammar of Motives.* New York: Braziller.
1973 *The Philosophy of Literary Form.* Berkeley: University of California Press.

Burling, R.
1964 Cognition and Componential Analysis: God's Truth or Hocuspocus? *American Anthropologist* 66:20–28.
Buttel, Frederick H.
1987 New Directions in Environmental Sociology. *Annual Review of Sociology* 13:465–68.
1992 Environmentalization: Origins, Processes, and Implications for Rural Social Change. *Rural Sociology* 57, no. 1:1–27.
Buttel, F. H., and W. L. Flinn
1974 The Structure of Support for the Environmental Movement, 1968–1970. *Rural Sociology* 39, no. 1:56–69.
Buttel, Frederick, and Peter Taylor
1992 Environmental Sociology and Global Environmental Change: A Critical Assessment. *Society and Natural Resources* 5:211–30.
Callicott, J. Baird
1994 *Earth's Insights: A Survey of Ecological Ethics from the Mediterranean Basin to the Australian Outback.* Berkeley: University of California Press.
Capra, F.
1991 *Belonging to the Universe: Explorations on the Frontiers of Science and Spirituality.* San Francisco: Harper.
Carman, J. B., and Frederique Appfel Marglin
1985 *Purity and Auspiciousness in Indian Society.* Leiden: E. J. Brill.
Central Pollution Control Board.
1993 *Annual Report.* New Delhi: Central Pollution Control Board.
1995a *Status of the Industrial Pollution Control Programme along the River Ganga (Phase-I).* New Delhi: Central Pollution Control Board.
1995b *Parivesh.* Newsletter for the Central Pollution Control Board 22, no. 2.
1995c *Pollution Control Acts, Rules and Notifications Issued Thereunder.* 3d ed. New Delhi: Central Pollution Control Board.
Chapple, Christopher Key
1998 Toward an Indigenous Indian Environmentalism. In *Purifying the Earthly Body of God: Religion and Ecology in Hindu India,* ed. Lance E. Nelson. Albany: State University of New York Press.
Chapple, Christopher Key, and Mary Evelyn Tucker, eds.
2000 *Hinduism and Ecology.* Cambridge: Harvard University Press.
Chaubey, B. B.
1970 *Treatment of Nature in the RGVeda.* Hoshiarpur: Vedic Sahitya Sadan.
Chaudhri, S. D., ed.
1904 *Criminal Law Journal of India,* 1.
Chaujar, R. K., R. K. Mazari, and J. T. Gergan
1993 Glacial Geomorphology of the Gaumukh: The Source of the Ganga, with Reference to Its Present State of Environment. Seminar on Ganga in the Service of the Nation. September 12–13.

Chinmayananda, Swami
1997 Commentary. In *The Holy Geeta.* Sri Ram Batra. Bombay: Central Chinmaya Mission Trust.
Clifford, James
1983 On Ethnographic Authority. *Representations* 1, no. 2 (spring):118–46.
1986 Introduction: Partial Truths. In *Writing Culture*, ed. James Clifford and George Marcus. Berkeley: University of California Press.
Commoner, Barry
1987 A Reporter at Large: The Environment. *New Yorker,* 15 June, 46–71.
Conklin, H. C.
1954 The Relation of Hanunoo Culture to the Plant World. Ph.D. diss., Yale University.
Costa, Alberto C. G., Conrad P. Kottak, Rosane M. Prado, and John Stiles
1994 Environmental Awareness and Risk Perception in Brazil. *Bulletin of the National Association of Practicing Anthropologists,* 71–87. Washington: American Anthropological Association.
Cremo, Michael A., and Mukunda Goswami
1995 *Divine Nature: A Spiritual Perspective on the Environmental Crisis.* Los Angeles: Bhaktivedanta Book Trust.
Crumley, Carole L.
1998 Foreword. In *Advances in Historical Ecology*, ed. William Balee. New York: Columbia University Press.
Curtis, Sue Ann
1992 Cultural Relativism and Risk Assessment Strategies for Federal Projects. *Human Organization* 51, no. 1:65–70.
Cvetkovich, George, and Timothy C. Earle
1992 Environmental Hazards and the Public. *Journal of Social Issues* 48, no. 4:1–20.
Dake, Karl
1992 Myths of Nature: Culture and the Social Construction of Risk. *Journal of Social Issues* 48, no. 4:21–37.
Daniel, E. Valentine
1984 *Fluid Signs: Being a Person the Tamil Way.* Berkeley: University of California Press.
Das, Veena
1977 *Structure and Cognition: Aspects of Hindu Caste and Ritual.* Delhi: Oxford University Press.
Das Gupta, S. P., ed.
1984 *Basin Sub-basin Inventory of Water Pollution: The Ganga Basin.* Part II. *Center for Study of Man and Environment.* Calcutta: Central Board for the Prevention and Control of Water Pollution.
Derman, Bill
1995 Environmental NGOs, Dispossession and the State: The Ideology and Praxis of African Nature and Development. *Human Ecology* 23, no. 2:199–216.
1998 Balancing the Waters: Development and Hydropolitics in Contemporary Zimbabwe. In *Water, Culture and Power: Local Struggles in a Global*

Context, ed. John M. Donahue and Barbara Rose Johnston, 73–94. Washington, DC: Island Press.

Derrida, Jacques

1976 *Of Grammatology*. Trans. Gayatri Spivak. Baltimore: Johns Hopkins University Press.

1978 *Writing and Difference*. Trans. Alan Bass. Chicago: University of Chicago Press.

Dimmit, C., and J. A. B. van Buitenen

1978 *Classical Hindu Mythology: A Reader in the Sanskrit Puranas*. Philadelphia: Temple University Press.

Dirks, Nicolas

1992 *Colonialism and Culture*. Ann Arbor: University of Michigan Press.

Donahue, John M., and Barbara Rose Johnston, eds.

1998 *Water, Culture and Power: Local Struggles in a Global Context*. Washington, DC: Island Press.

Douglas, M.

1966 *Purity and Danger: An Analysis of the Concepts of Pollution and Taboo*. London: Routledge and Kegan Paul.

1985 *Risk Acceptability According to the Social Sciences*. New York: Russel Sage Foundation.

1992 Environments at Risk. In *Ecology: The Shaping Enquiry*, ed. J. Benthall, 129–45. London: Longman.

Douglas, Mary, and Aaron Wildavsky

1982 *Risk and Culture*. Berkeley: University of California Press.

Downey, Gary L.

1986 Risk in Culture: The American Conflict over Nuclear Power. *Cultural Anthropology* 1:388–412.

Dumont, Louis

1980 *Homo Hierarchicus*. 3d ed. Chicago: University of Chicago Press.

Dumont, Louis, and D. F. Pocock

1959 Pure and Impure. *Contributions to Indian Sociology* 3:9–34.

Eck, Diana

1980 A Survey of Sanskrit Sources for the Study of Varanasi. *Purana* 22, no. 1 (January): 81–101.

1981 *Darsan: Seeing the Divine Image in India*. Chambersburg: Anima.

1982a *Banaras: City of Light*. New York: Alfred Knopf.

1982b Ganga: The Goddess in Hindu Sacred Geography. In *The Divine Consort: Radha and the Goddesses of India*, ed. John Stratton Hawley and Donna Mane Wulff, 166–93. Boston: Beacon Press.

Eder, Klaus

1996 The Institutionalisation of Environmentalism: Ecological Discourse and the Second Transformation of the Public Sphere. In *Risk, Environment and Modernity: Towards a New Ecology*, ed. Scott Lash, Bronislaw Szerszynski, and Brian Wynne, 203–23. London: Sage.

Ellen, Roy

1982 *Environment, Subsistence and System: The Ecology of Small-Scale Social Formations*. Cambridge University Press.

Ellen, Roy F., and David Reason
1979 *Classifications in Their Social Context.* London: Academic.
Escobar, Arturo
1995 *Encountering Development: The Making and Unmaking of the Third World.* Princeton: Princeton University Press.
1996 Construction Nature: Elements for a Post-structuralist Political Ecology. *Futures* 28, no. 4:325–43.
1999 After Nature: Steps to an Antiessentialist Political Ecology. *Current Anthropology* 40, no. 1:1–30.
Escobar, Arturo, and Sonia E. Alvarez
1992 *The Making of Social Movements in Latin America.* Boulder: Westview.
Ettenger, Kreg
1998 "A River That Was Once So Strong and Deep": Local Reflections on the Eastmain Diversion, James Bay Hydroelectric Project. In *Water, Culture and Power: Local Struggles in a Global Context,* ed. John M. Donahue and Barbara Rose Johnston, 47–72. Washington, DC: Island Press.
Feeny, David, Fikret Berkes, Bonnie McCay, and James Acheson
1992 The Tragedy of the Commons: Twenty-two Years Later. *Human Ecology* 18, no. 1:1–19.
Feldhaus, Anne
1995 *Water and Womanhood.* New York: Oxford University Press.
Ferguson, James
1990 *The Anti-Politics Machine.* Cambridge: Cambridge University Press.
Fernandez, James
1974 The Mission of Metaphor in Expressive Culture. *Current Anthropology* 15, no. 2:119–45.
1977 The Performance of Ritual Metaphors. In *The Social Use of Metaphors,* ed. J. D. Sapir and J. C. Crocker, 100–131. Philadelphia: University of Pennsylvania Press.
1998 Trees of Knowledge of Self and Other in Culture: On Models for the Moral Imagination. In *The Social Life of Trees,* ed. L. Rival, 81–110. Oxford: Berg.
Fitchen, Janet M.
1987 Cultural Aspects of Environmental Problems: Individualism and Chemical Contamination of Groundwater. *Science, Technology and Human Values* 12, no. 2:1–12.
1989 When Toxic Chemicals Pollute Residential Environments: The Cultural Meanings of Home and Homeownership. *Human Organization* 48, no. 4:313–24.
Fitzjames, F.
1880 *Preliminary Report on the Sewerage and Water Supply of the City of Benaras.* Government of NW Provinces and Oudh, Public Works Dept. Allahabad: North-Western Provinces and Oudh Press.
Foucault, Michel
1972a *Power/Knowledge.* Pantheon.

1972b *The Archaeology of Knowledge.* Trans. A. M. Sheridan Smith. New York: Harper Colophon.
1975 *I, Pierre Rivière, having slaughtered my mother, my sister, and my brother . . . A Case of Parricide in the Nineteenth Century.* Trans. Frank Jellinek. Lincoln: University of Nebraska Press.
1980 *History of Sexuality.* Vol. I. *An Introduction.* Trans. Robert Hurley. New York: Vintage/Random House.
1991 Governmentality. In *The Foucault Effect*, ed. Graham Burchell, Colin Gordon, and Peter Miller, 87–104. Chicago: University of Chicago Press.

Fuller, C. J.
1979 Gods, Priests and Purity: On the Relation between Hinduism and the Caste System. *Man*, n.s., 14:459–76.
1992 *The Camphor Flame.* Princeton: Princeton University Press.

Gadgil, Madhav, and Ramachandra Guha
1992 *This Fissured Land.* Berkeley: University of California Press.

Galanter, Marc
1989 *Law and Society in Modern India.* Delhi: Oxford University Press.

Geertz, Clifford
1973 Deep Play: Notes on a Balinese Cockfight. In *The Interpretation of Cultures*, 412–53. New York: Basic Books.

Gerlach, Luther P.
1987 Protest Movements and the Construction of Risk. In *The Social and Cultural Construction of Risk*, ed. B. Johnson and Covello, 103–45. Dordrecht: Reidel.

Gerlach, Luther P., and Larry R. Meiller
1987 Social and Political Process: The Minnesota Hazardous Waste Case. In *Needs Assessment: Theory and Methods*, ed. Donald E. Johnson, Larry R. Meiller, Lorna Clancy Miller, and Gene F. Summers. Ames: Iowa State University Press.

Gerlach, Luther P. and S. Rayner
1988 Culture and the Common Management of Global Risks. *Practicing Anthropology* 10 (3–4): 15–18.

Gold, Ann
1998 Sin and Rain: Moral Ecology in Rural North India. In *Purifying the Earthly Body of God: Religion and Ecology in Hindu India*, ed. Lance E. Nelson, 165–95. Albany: State University of New York Press.

Gondriaan, Teun
1978 *Māyā Human and Divine.* Delhi: Motilal Banarsidass.

Government of India
1965 *Uttar Pradesh District Gazetteer.*

Greaves, Tom
1998 Rights in the Pacific Northwest. In *Water, Culture and Power: Local Struggles in a Global Context*, ed. John M. Donahue and Barbara Rose Johnston. Washington, DC: Island Press.

Grossman, Karl

1993 Environmental Racism. In *The Racial Economy of Science: Toward a Democratic Future*, ed. Sandra Harding, 326–34. Bloomington: Indiana University Press.

Guha, Ramachandra

1983 Forestry in British and Post-British India: A Historical Analysis. *Economic and Political Weekly*, 29 October, 1882.

1994 Colonialism and Conflict in the Himalayan Forest. In *Social Ecology*, ed. Ramachandra Guha. Delhi: Oxford University Press.

Guha, Ranajit, ed.

1982 *Subaltern Studies: Writings on South Asian History and Society.* Vol. 6. Delhi: Oxford University Press.

Guha, Ranajit, and Gayatri Spivak, eds.

1988 *Selected Subaltern Studies.* Delhi: Oxford University Press.

Gupta, A.

1995 Blurred Boundaries: The Discourses of Corruption, the Culture of Politics and the Imagined State. *American Ethnologist* 22, no. 2:375–402.

1998 *Postcolonial Developments.* Stanford: Stanford University Press.

Gupta, A., and James Ferguson

1997a Culture, Power, Place: Ethnography at the End of an Era. In *Culture, Power, Place: Explorations in Critical Anthropology*, ed. Akhil Gupta and James Ferguson. Durham: Duke University Press.

1997b Beyond Culture: Space, Identity and the Politics of Difference. In *Culture, Power, Place: Explorations in Critical Anthropology*, ed. Akhil Gupta and James Ferguson. Durham: Duke University Press.

1997c Discipline and Practice: "The Field" as Site, Method, and Location in Anthropology. In *Anthropological Locations*, ed. A. Gupta and J. Ferguson. Berkeley: University of California Press.

Haas, Peter

1992 Introduction: Epistemic Communities and International Policy Coordination. *International Organization* 46, no. 1:1–35.

Hallman, William, and Abraham Wandersman

1992 Attribution of Responsibility and Individual Collective Coping with Environmental Threats. *Journal of Social Issues* 48, no. 4:101–18.

Hamilton, Walter

1993 *East-India Gazetteer.* 2d ed. Vol. I. Delhi: Low Price Publications.

Hammad, S. Md.

1992 Development of Varanasi Sewerage System and Prevention of Pollution to River Ganga. In *A Seminar on Pollution Control of River Cities of India: A Case Study of Varanasi.* Varanasi: Sankat Mochan Foundation.

Handelman, Don, and Lea Shamgar-Handelman

1990 Shaping Time: The Choice of the National Emblem of Israel. In *Culture through Time*, ed. Emiko Ohnuki-Tierney, 193–226. Stanford: Stanford University Press.

Hansen, Thomas Blom
1999 *The Saffron Wave: Democracy and Hindu Nationalism in Modern India.* Princeton: Princeton University Press.
Hargrove, Eugene
1986 *Religion and Environmental Crisis.* Athens: University of Georgia Press.
Harper, Edward B.
1964 Ritual Pollution as an Integrator of Caste and Religion. *Journal of Asian Studies* 23:151–97.
Harris, Marvin
1968 *The Rise of Anthropological Theory.* London: Routledge and Kegan Paul.
Havell, E. B.
1990 *Benares: The Sacred City.* Varanasi: Vishwavidyalaya Prakashan. [1905].
Hayles, N. Katherine
1995 Searching for Common Ground. In *Reinventing Nature? Responses to Postmodern Deconstruction,* ed. Michael E. Soula and Gary Lease. Washington, DC: Island Press.
Haynes, D., and G. Prakash, eds.
1992 *Contesting Power: Resistance and Everyday Social Relations in South Asia.* Berkeley: University of California Press.
Hershkovitz, L.
1993 Political Ecology and Environmental Management in the Loess Plateau, China. *Human Ecology* 21:327–53.
Herzfeld, Michael
1993 *The Social Production of Indifference: Exploring the Symbolic Roots of Western Bureaucracy.* Chicago: University of Chicago Press.
Hunter, W. W.
1888 *The Imperial Gazetteer of India.* Vol. IV. London: Trubner.
Hussain, Aizaz
1989 *Whither Local Bodies in India.* Delhi: Criterion.
Inden, R.
1976 *Marriage and Rank in Bengali Culture.* Berkeley: University of California Press.
India International Centre
1992 *Indigenous Vision: Peoples of India, Attitudes to the Environment.* India International Centre Quarterly. Delhi: India International Centre.
Indian Law Reports
1876
[Indian] *Penal Code, 1860.*
1860 Air Man. Vol. 28.
Ingold, Tim
1993 Globes and Spheres: The Topology of Environmentalism. In *Environmentalism: The View from Anthropology,* ed. Kay Milton, 31–42. London: Routledge.

Jameson, Fredric
1998 Notes on Globalization as a Philosophical Issue. In *The Cultures of Globalization*, ed. Fredric Jameson and Masao Miyoshi, 54–77. Durham: Duke University Press.

Jariwala, C. M.
2000 The Directions of Environmental Justice: An Overview. In *Fifty Years of the Supreme Court of India*, ed. S. K. Verma and Kusum, 469–94. Delhi: Oxford University Press.

Jha, Ajit Kumar
1995 Hindutva Swadeshi: Waves of Economic Nationalism. *Times of India* (New Delhi), Editorial. 6 September.

Kane, P. V.
1973 *History of Dharmasastra*. 2d ed. Vol IV. Government Oriental Series Class B. No. 6. Poona: Bhandarkar Oriental Research Institute.

Kaniaru, Donald, Lal Kurukulasuriya, Prasantha Dias Abeyegunawardene, and Chad Martino
1997 *Report of the Regional Symposium on the Role of the Judiciary in Promoting the Rule of Law in the Area of Sustainable Development*. South Asia Co-operative Environment Programme and United Nations Environment Programme.

Kasperson, Roger E., Dominic Golding, and Seth Tuler
1992 Social Distrust as a Factor in Siting Hazardous Facilities and Communicating Risks. *Journal of Social Issues* 48, no. 4:161–87.

Kay, Paul
1971 Taxonomy and Semantic Contrast. *Language* 68:866–87.

Kempton, Willett, James S. Boster, and Jennifer A. Hartley
1995 *Environmental Values in America*. Cambridge: MIT Press.

Kinsley, David
1987 *Hindu Goddesses: Visions of the Divine Feminine in the Hindu Religious Tradition*. Delhi: Motilal Banarsidass.

Korom, Frank
1998 On the Ethics and Aesthetics of Recycling in India. In *Purifying the Earthly Body of God: Religion and Ecology in Hindu India*, ed. Lance E. Nelson. Albany: State University of New York Press.

Kottak, Conrad
1999 The New Ecological Anthropology. *American Anthropologist* 101, no. 1:23–35.

Kottak, Conrad P., and Alberto C. G. Costa
1993 Ecological Awareness, Environmentalist Action and International Conservation Strategy. *Human Organization* 52, no. 4:335–43.

Kumar, Nita
1988 *The Artisans of Banaras*. Princeton: Princeton University Press.

Lansing, J. Stephen
1991 *Priests and Programmers: Technologies of Power in the Engineered Landscape of Bali*. Princeton: Princeton University Press.

Leach, Edmund

1990 Aryan Invasions over Four Millennia. In *Culture through Time*, ed. Emiko Ohnuki-Tierney, 227–45. Stanford: Stanford University Press.

Lease, Gary

1995 Introduction: Nature under Fire. In *Reinventing Nature? Responses to Postmodern Deconstruction*, ed. Michael E. Soula and Gary Lease. Washington, DC: Island Press.

Lingat, Robert

1973 *The Classical Law of India*. Berkeley: University of California Press.

Loker, William M.

1998 Water, Rights, and El Cajon Dam, Honduras. In *Water, Culture and Power: Local Struggles in a Global Context*, ed. John M. Donahue and Barbara Rose Johnston, 95–119. Washington, DC: Island Press.

Lovelock, James

1988 *The Ages of Gaia: A Biography of Our Loving Earth*. New York: Norton.

Low, Setha

1996 The Anthropology of Cities: Imagining and Theorizing the City. *Annual Review of Anthropology* 25:383–409.

Ludden, David

1993 Orientalist Empiricism: Transformations of Colonial Knowledge. In *Orientalism and the Post-Colonial Predicament*, ed. Carol Breckenridge and Peter van der Veer. Philadelphia: University of Pennsylvania Press.

Luke, Timothy W.

1995 On Environmentality: Geo-Power and Eco-Knowledge in the Discourses of Contemporary Environmentalism. *Cultural Critique* 31:57–81.

Lutgendorf, P.

1991 *The Life of a Text*. Berkeley: University of California Press.

Lutz, Catherine, and Geoffrey M. White

1986 The Anthropology of Emotions. *Annual Review of Anthropology* 15:405–36.

Lynch, Owen, ed.

1990 *Divine Passions: The Social Construction of Emotion in India*. Berkeley: University of California Press.

1996 Contesting and Contested Identities: Mathura's Chaubes. In *Narratives of Agency: Self-making in China, India and Japan*, ed. W. Dissanayake, 74–103. Minneapolis: University of Minnesota Press.

Madan, T. N.

1985 Concerning the Categories Subha and Suddha in Hindu Culture: An Exploratory Essay. In *Purity and Auspiciousness in Indian Society*, ed. John B. Carman and Frederique Appfel-Marglin. Leiden: E. J. Brill.

1987 Secularism in Its Place. *Journal of Asian Studies* 46, no. 4:747–59.

1993 Whither Indian Secularism? *Modern Asian Studies* 27, no. 3:667–97.

Maffi, Luisa, ed.
2001 *On Biocultural Diversity: Linking Language, Knowledge, and the Environment.* Washington, DC: Smithsonian Institution Press.
Malik, Surendra
1988 *Supreme Court Yearly Digest.* Lucknow: Eastern Book Co.
Malik, Surendra, and P. L. Malik, eds.
1991 *The Supreme Court Cases 1991.* Supplement Vol. 1. Lucknow: Eastern Book Company.
1992 *The Supreme Court Cases 1992.* Supplement Vol. 2. Lucknow: Eastern Book Company.
1993 *The Supreme Court Cases 1993.* Supplement Vol. 1. Lucknow: Eastern Book Company.
Manohar, V. R., and W. W. Chitaley
1979a *The A.I.R. Manual.* 4th ed. Vol. 19. Nagpur: All India Reporter.
1979b *The A.I.R. Manual.* 4th ed. Vol. 27. Nagpur: All India Reporter.
1979c *The A.I.R. Manual.* 4th ed. Vol. 28. Nagpur: All India Reporter.
1979d *The A.I.R. Manual.* 4th ed. Vol. 31. Nagpur: All India Reporter.
Marcus, George
1986 Contemporary Problems of Ethnography in the Modern World System. In *Anthropology as Cultural Critique*, eds. James Clifford and George E. Marcus, 165–93. Chicago: University of Chicago Press.
1995 Ethnography in/of the World System: The Emergence of Multi-sited Ethnography. *Annual Review of Anthropology* 24:95–117.
Marcus, George, and Michael Fischer
1986 *Anthropology as Cultural Critique: An Experimental Movement in the Human Sciences.* Chicago: University of Chicago Press.
Marglin, Frederique Apffel
1977 Power, Purity and Pollution: Aspects of the Caste System Reconsidered. *Contributions to Indian Sociology* 2, no. 2:245–70.
1985a Types of Oppositions in Hindu Culture. In *Purity and Auspiciousness in Indian Society*, ed. John B. Carman and F. A. Marglin. Leiden: E. J. Brill.
1985b Purity, Power and Pollution. In *Purity and Auspiciousness in Indian Society*, ed. John B. Carman and F. A. Marglin. Leiden: E. J. Brill.
Marriott, McKim
1989 Constructing an Indian Ethnosociology. *Contributions to Indian Sociology* 23, no. 1:1–39.
Marriott, McKim, and R. Inden
1977 Toward an Ethnosociology of South Asian Caste Systems. In *The New Wind: Changing Identities in South Asia*, ed. Kenneth David, 423–28. Chicago: Aldine.
Martinez-Alier, Joan
1998 'Environmental Justice' (Local and Global). In *The Cultures of Globalization*, ed. Fredric Jameson and Masao Miyoshi, 312–26. Durham: Duke University Press.

Mathur, L. N.
1980 A Federal Legislative History of Control of Water Pollution in India. In *Legal Control of Environmental Pollution*, ed. S. Agarwal. Bombay: N. M. Tripathi Pvt Ltd.

McCarthy, John D., and Mayer N. Zald
1973 *The Trend in Social Movements in America: Professionalism and Resource Mobilization.* Morristown: General Learning Press.
1977 Resource Mobilization in Social Movements: A Partial Theory. *American Journal of Sociology* 82:1212–39.

McCay, B., and J. Acheson, eds.
1987 *The Question of the Commons: The Culture and Ecology of Communal Resources.* Tucson: University of Arizona.

McGrath, D., F. de Castro, C. Futemma, B. D. de Amaral, and J. Calabria
1993 Fisheries and the Evolution of Resource Management on the Lower Amazon Floodplain. *Human Ecology* 21, no. 2:167–95.

Mehta, M. C.
n.d. Law and Jurisprudence for Environment Protection and Sustainable Development in South Asia—With Special Focus on India. Unpublished paper.

Miller, Barbara Stoler
1991 Presidential Address: Contending Narratives—The Political Life of the Indian Epics. *Journal of Asian Studies* 50, no. 4:783–92.

Miller, Vernice D.
1993 Building on Our Past, Planning for Our Future: Communities of Color and the Quest for Environmental Justice. In *Toxic Struggles: The Theory and Practice of Environmental Justice*, ed. R. Hofrichter. Philadelphia: New Society.

Miller, Vernice, Moya Hallstein, and Susan Quass
1996 Feminist Politics and Environmental Justice. In *Feminist Political Ecology: Global Issues and Local Experiences*, ed. Dianne Rocheleau, Barbara Thomas-Slayter, and Esther Wangari, 62–85. New York: Routledge.

Ministry of Environment and Forests
1993 *Ganga Action Plan Achievements.* Project Directorate. Central Ganga Authority. Ministry of Environment and Forests. Pamphlet.
1994 *National River Action Plan.* April. New Delhi: Government of India.
1995 *Evaluation of Ganga Action Plan.* April. New Delhi: Government of India.
1998 *Status Paper on the River Action Plans.* New Delhi: Ministry of Environment and Forests.

Mitra, Subrata Kumar
1991 Desecularizing the State: Religion and Politics in India after Independence. *Comparative Studies in Society and History* 33, no. 4:775–77.

Monteiro, John B.
1966 *Corruption: Control of Maladministration.* Bombay: Manaktalas.

Moran, Emilio F.
1990 *The Ecosystem Approach in Anthropology.* Ann Arbor: University of Michigan Press.

Mumme, Patricia
1998 Models and Images for a Vaisnava Environmental Theology: The Potential Contribution of Srivaisnavism. In *Purifying the Earthly Body of God: Religion and Ecology in Hindu India*, ed. Lance E. Nelson, 133–61. Albany: State University of New York Press.

Murti, C. R. K., K. S. Bilgrami, T. M. Das, and R. P. Mathur, eds.
1991 *The Ganga: A Scientific Study*. Ganga Project Directorate. New Delhi: Ganga Project Directorate.

Murty, M. N.
1995 Use of Economic Instruments for Controlling Pollution in India. Working Paper Series. Delhi: Institute of Economic Growth.

Nagarajan, Vijaya Rettakudi
1998 The Earth as Goddess Bhu Devi: Toward a Theory of "Embedded Ecologies" in Folk Hinduism. In *Purifying the Earthly Body of God: Religion and Ecology in Hindu India*, ed. Lance E. Nelson, 269–96. Albany: State University of New York Press.

Nandy, Ashis
1983 *The Intimate Enemy: Loss and Recovery of Self under Colonialism*. Delhi: Oxford University Press.
1995 History's Forgotten Doubles. *History and Theory* 34, no. 2:44–67.

Narain, V. A.
1959 *Jonathan Duncan and Varanasi*. Calcutta.

Narayanan, Vasudah
2001 Water, Wood, and Wisdom: Ecological Perspectives from the Hindu Traditions. *Daedalus*, fall.

Nath, Pashupati, and Siddha Nath
1990 *Environmental Pollution*. Allahabad: Chugh Publications.

Nelson, Lance E., ed.
1998 *Purifying the Earthly Body of God: Religion and Ecology in Hindu India*. Albany: State University of New York Press.

Novotny, Patrick
2000 *Where We Live, Work and Play: The Environmental Justice Movement and the Struggle for a New Environmentalism*. London: Praeger.

O'Connor, James
1988 Capitalism, Nature, Socialism: A Theoretical Introduction. *Capitalism, Nature, Socialism* 1, no. 1:11–38.
1989 Political Economy of Ecology of Socialism and Capitalism. *Capitalism, Nature, Socialism* 1, no. 3:93–108.
1992 A Political Strategy for Ecology Movements. *Capitalism, Nature, Socialism* 3, no. 1:1–5.

O'Connor, Martin
1993 On the Misadventures of Capitalist Nature. *Capitalism, Nature, Socialism* 4, no. 3:7–40.

O'Flaherty, W. D.
1976 *The Origins of Evil in Hindu Mythology*. Berkeley: University of California Press.
1981 *The Rig Veda: An Anthology*. London: Penguin.

Ohnuki-Tierney, Emiko

1981 Phases in Human Perception/Cognition/Symbolization Processes: Cognitive Anthropology and Symbolic Classification. *American Ethnologist* 8, no. 3:451–67.

1990a Introduction: The Historicization of Anthropology. In *Culture through Time*, ed. Emiko Ohnuki-Tierney, 1–25. Stanford: Stanford University Press.

1990b The Monkey as Self in Japanese Culture. In *Culture through Time*, ed. Emiko Ohnuki-Tierney, 128–53. Stanford: Stanford University Press.

Oldenburg, Philip

1987 Middlemen in Third World Corruption: Implications of an Indian Case. *World Politics* 39 (July):508–35.

Orlove, Benjamin S.

1991 Mapping Reeds and Reading Maps: The Politics of Representation in Lake Titicaca. *American Ethnologist* 18, no. 1:3–38.

Pandit, Ajay, ed.

1989 *Annual Supreme Court Digest 1988*. Delhi: Judgements Today.

Parikh, Manju

1993 The Debacle at Ayodhya. *Asian Survey* 33, no. 7:673–84.

Parkin, David

1985 Entitling Evil: Muslims and Non-Muslims in Coastal Kenya. In *The Anthropology of Evil*, ed. David Parkin. Oxford: Basil Blackwell.

Parmanand

1985 *Mahamana Madan Mohan Malaviya*. Varanasi: Malaviya Adhyayan Sansthan.

Parry, Jonathon

1980 Ghosts, Greed and Sin: the Occupational Identity of the Benaras Funeral Priests. *Man*, n.s., 15, no. 1:88–11.

1981 Death and Cosmogony in Kashi. *Contributions to Indian Sociology* 15, no. 1–2:337–65.

1994 *Death in Banaras*. Cambridge: Cambridge University Press.

Paul, Samuel

1997 Corruption: Who Will Bell the Cat? *Economic and Political Weekly* 32, no. 23:1350–55.

Peet, Richard, and Michael Watts

1996 *Liberation Ecologies: Environment, Development, Social Movements*. New York: Routledge.

Peluso, Nancy

1992 *Rich Forests, Poor People: Resource Control and Resistance in Java*. Berkeley: University of California Press.

People's Commission on Environment and Development.

1994 *Report: Public Hearing on Environment and Development*. Varanasi. 8 November.

Picchi, Debra

1995 Village Division in Lowland South America: The Case of the Bakairi Indians of Central Brazil. *Human Ecology* 23, no. 4:477–98.

Pigg, Stacy Leigh

1992 Constructing Social Categories through Place: Social Representations and Development in Nepal. *Comparative Studies in Society and History* 23, no. 4:565–90.

Potter, David C.

1996 *India's Political Administrations: From ICS to IAS.* Delhi: Oxford.

Raheja, G.

1988 *The Poison in the Gift: Ritual, Prestation and the Dominant Caste in a North Indian Village.* Chicago: University of Chicago Press.

Rangan, Haripriya

1996 From Chipko to Uttaranchal: Development, Environment and Social Protest in the Garhwal Himalayas, India. In *Liberation Ecologies: Environment, Development, Social Movements,* ed. Richard Peet and Michael Watts, 205–26. New York: Routledge.

Rappaport, Roy A.

1967 *Pigs for the Ancestors: Ritual in the Ecology of a New Guinea People.* New Haven: Yale University Press.

1984 *Pigs for the Ancestors: Ritual in the Ecology of a New Guinea People.* Enlarged ed. New Haven: Yale University Press.

1990 Ecosystems, Populations and People. In *The Ecosystem Approach in Anthropology,* ed. Emilio F. Moran, 41–72. Ann Arbor: University of Michigan Press.

Rees, William, and Mathis Wackernagel

1994 Ecological Footprints and Appropriated Carrying Capacity. In *Investing in Natural Capital: The Ecological Economics Approach to Sustainability,* ed. A. M. Jansson, Monica Hammer, et al. Washington, DC: Island Press.

Reisner, Marc

1986 *Cadillac Desert: The American West and Its Disappearing Water.* New York: Penguin.

1990 *Overtapped Oasis: Reform or Revolution for Western Water.* Washington, DC: Island Press.

Rocheleau, Dianne, Barbara Thomas-Slayter, and Esther Wangari, eds.

1996 *Feminist Political Ecology: Global Issues and Local Experiences.* New York: Routledge.

Rocher, Rosane

1993 British Orientalism in the Eighteenth Century: The Dialectics of Knowledge and Government. In *Orientalism and the Postcolonial Predicament,* ed. Carol Breckenridge and Peter van der Veer, 215–49. Philadelphia: University of Pennsylvania Press.

Rogers, Claudia M.

1998 Water Resource Development and Its Effects on the Human Community: The Tennessee-Tombigbee Waterway, Southeastern United States. In *Water, Culture and Power: Local Struggles in a Global Context,* ed. John M. Donahue and Barbara Rose Johnston, 123–40. Washington, DC: Island Press.

Rosaldo, Michelle Z.
1980 *Knowledge and Passion: Ilongot Notions of Self and Social Life.* Cambridge: Cambridge University Press.
1984 Toward an Anthropology of Self and Feeling. In *Culture Theory: Essays on Mind, Self, and Emotion*, ed. Richard A. Shweder and Robert A. LeVine, 137–57. Cambridge: Cambridge University Press.
Rosaldo, Renato
1989 *Culture and Truth: The Remaking of Social Analysis.* Boston: Beacon.
Rosencranz, Armin, Shyam Divan, and Martha L. Noble
1991 *Environmental Law and Policy in India: Cases, Materials, and Statutes.* Bombay: Tripathi.
Rosenthal, D. B.
1970 *The Limited Elite: Politics and Government in Two Indian Cities.* Chicago: University of Chicago Press.
Sahlins, Marshall
1976 *Culture and Practical Reason.* Chicago: University of Chicago Press.
Said, Edward
1978 *Orientalism.* New York: Vintage Books.
Saldanha, Michael F.
1998 People's Initiatives and Judicial Activism as a Catalyst of Institutional Reform. *Fifth International Conference on Environmental Compliance and Enforcement.* Conference Proceedings vol. 1. 16–20 November.
Sale, Kirkpatrick
1986 The Forest for the Trees: Can Today's Environmentalists Tell the Difference? *Mother Jones* 11, no. 8:25–33.
Sankat Mochan Foundation
1990 *Swatchha Ganga Campaign Annual Report 1988–1990.* Varanasi.
1992 *A Seminar on Pollution Control of River Cities in India: A Case Study of Varanasi.* Varanasi.
1994 *Proposal for GAP Phase II at Varanasi.* Varanasi: Swatcha Ganga Campaign.
Saussure, Ferdinand de
1966 *Course in General Linguistics.* New York: McGraw-Hill.
Schjolden, Ane
2000 *Leather Tanning in India: Environmental Regulations and Firms' Compliance.* F-I-L Working Papers, no. 21.
Schmink, Marianne, and Charles H. Wood
1987 The "Political Ecology" of Amazonia. In *Lands at Risk in the Third World*, ed. P. Little, Michael Horowitz, and A. Nyerges, 38–57. Boulder: Westview Press.
1992 *Contested Frontiers in Amazonia.* New York: Columbia University Press.
Schnaiberg, Allan
1973 Politics, Participation and Pollution: The Environmental Movement. In *Cities in Change: Studies on the Urban Condition*, ed. John Walton and Donald Carns, 605–27. Boston: Allyn and Bacon.

1980 *The Environment: From Surplus to Scarcity.* New York: Oxford University Press.
1985 The Retreat from Political to Technical Environmentalism. In *Social Responses to Technological Change: Contributions to Sociology,* 56, ed. Augustine Brannigan and Sheldon Goldenburg, 19–36. Westport: Greenwood.

Scott, James
1985 *Weapons of the Weak: Everyday Forms of Peasant Resistance.* New Haven: Yale University Press.
1990 *Domination and the Arts of Resistance: Hidden Transcripts.* New Haven: Yale University Press.

Sharan, R. K., and R. K. Sinha
1988 *Ganga Basin Research Project (Buxar-Barh).* Ganga Project Directorate Project. Patna: Patna University.

Sharma, Anju
1998 Climate No Headway. *Down to Earth* 17, no. 14 (December 15).

Sheridan, Thomas E.
1998 The Big Canal: The Political Ecology of the Central Arizona Project. In *Water, Culture and Power: Local Struggles in a Global Context,* ed. John M. Donahue and Barbara Rose Johnston, 163–86. Washington, DC: Island Press.

Sherma, Rita DasGupta
1998 Sacred Immanence: Reflections of Ecofeminism in Hindu Tantra. In *Purifying the Earthly Body of God: Religion and Ecology in Hindu India,* ed. Lance E. Nelson, 89–131. Albany: State University of New York Press.

Shiva, Vandana
1989 *Staying Alive: Women, Ecology and Development.* London: Zed.
1992 Women's Indigenous Knowledge and Biodiversity Conservation. In *Indigenous Vision: Peoples of India, Attitudes to the Environment,* ed. India International Centre, 205–14. India International Centre Quarterly. Delhi: India International Centre.

Shkilnyk, Anastasia M.
1985 *A Poison Stronger than Love: The Destruction of an Ojibwa Community.* New Haven: Yale University Press.

Shukla, A. C., and A. Vandana
1995 *Ganga: A Water Marvel.* New Delhi: Ashish.

Singh, Gayatri, Kerban Anklesaria, and Colin Gonslaves, eds.
n.d. *The Environmental Activists' Handbook: Statute, Judgments, Strategies.* Bombay: Colin Gonsalves.

Singh, Gurdip
1995 *Environmental Law: International and National Perspectives.* New Delhi: Lawman Pvt. Ltd.

Singh, R. P. B., ed.
1993a *Banaras (Varanasi): Cosmic Order, Sacred City, Hindu Traditions.* Varanasi: Tara Book Agency.
1993b *Environmental Ethics.* Varanasi: National Geographic Society.

Sivaramamurti, C.
1976 *Ganga*. Delhi: Orient Longman.
Smelser, Neil
1963 *Theory of Collective Behavior*. New York: Free Press.
Soule, Michael E., and Gary Lease, eds.
1995 *Reinventing Nature? Responses to Postmodern Deconstruction*. Washington, DC: Island Press.
Srinivas, M. N.
1952 *Religion and Society among the Coorgs of South India*. Bombay: Asia.
Srivastava, Om Prie
1980 *Municipal Government and Administration in India*. Allahabad: Chugh.
Stallings, Robert
1973 Patterns of Belief in Social Movements: Clarifications from an Analysis of Environmentalist Groups. *Sociological Quarterly* 14:465–80.
Standing Manual for River Ganga and Ganges Canal. Uttar Pradesh Irrigation Department.
Stonich, Susan
1993 *I Am Destroying the Land! The Political Ecology of Poverty and Environmental Destruction in Honduras*. Boulder: Westview.
Subrahmaniam, Vidya
1995 Redefining Secularism: Gap between Theory and Practice. *Times of India* (New Delhi), 22 September.
Supreme Court of India Record of Proceedings. Writ Petition (C) No. 3727 of 1985 (7/14/95; 7/20/95; 10/20/95).
Szasz, Andrew
1994 *Ecopopulism: Toxic Waste and the Movement for Environmental Justice*. Social Movements, Protest and Contention, vol. 1. Minneapolis: University of Minnesota Press.
Thakur, Ramesh
1993 Ayodhya and the Politics of India's Secularism. *Asian Survey* 33 (July):645–64.
Thapar, Romila
1973 *Asoka and the Decline of the Mauryas*. Oxford: Oxford University Press.
Tilly, Charles
1978 *From Mobilization to Revolution*. Reading: Addison-Wesley.
Tiwari, I. C., and P. C. Sen
1991 The Pathogenic Bacteria in the Ganga Water Collected from Twelve Sampling Points in Varanasi. In *The Ganga*, ed. C. R. Krishna Murti, K. S. Bilgrami, T. M. Das, and R. P. Mathur, 78. New Delhi: Northern Book Centre.
Tripathi, B. D.
1991 An Overview of the Hydrobiological Features of the Ganga in the Stretch Mirzapur to Ballia. In *The Ganga*, ed. C. R. Krishna Murti, K. S. Bilgrami, T. M. Das, and R. P. Mathur, 156–60. New Delhi: Northern Book Centre.
Tsing, Anna Lowenhaupt
1993 *In the Realm of the Diamond Queen*. Princeton: Princeton University Press.

Tucker, Mary Evelyn, and John A. Grim
2001 Introduction: The Emerging Alliance of World Religious and Ecology. *Daedalus*, fall.

Turner, Ralph
1981 Collective Behavior and Resource Mobilization as Approaches to Social Movements. In *Research in Social Movements, Conflict and Change*. Vol. 4, ed. Louis Kreisberg. Greenwich: JAI.

Turner, Ralph, and Lewis M. Killian
1972 *Collective Behavior*. 2d ed. Englewood Cliffs: Prentice-Hall.

Turner, Victor
1975 *Dramas, Fields and Social Metaphors: Symbolic Action in Human Society*. Ithaca: Cornell University Press.

United Church of Christ Commission for Racial Justice
1987 *Toxic Wastes and Race in the United States: A National Study of the Racial and Socioeconomic Characteristics of Communities with Hazardous Waste Sites*. New York: United Church of Christ Commission for Racial Justice.

Upadhyaya, Prakash Chaudra
1992 The Politics of Indian Secularism. *Modern Asian Studies* 26, no. 4:815–53.

Uttar Pradesh Jal Nigam
1991 *Ganga Action Plan—Varanasi*. Varanasi: Ganga Pollution Prevention Unit.

Valerio, Valeri
1990 Constitutive History: Genealogy and Narrative in the Legitimation of Hawaiian Kingship. In *Culture through Time*, ed. Emiko Ohnuki-Tierney, 154–92. Stanford: Stanford University Press.

van Buitenen, J. A. B.
1973 *The Mahabharata*. Chicago: University of Chicago Press.

van der Veer, P.
1987 "God Must be Liberated!" A Hindu Liberation Movement in Ayodhya. *Modern Asian Studies* 21, no. 2:283–301.
1988 *Gods on Earth: The Management of Religious Experience and Identity in a North Indian Pilgrimage Center*. London: Athlo.
1994 *Religious Nationalism: Hindus and Muslims in India*. Berkeley: University of California Press.
1996 The Ruined Center: Religion and Mass Politics in India. *Journal of International Affairs* 50, no. 1:254–77.

Van Dyke, Virginia
1997 General Elections, 1996: Political Sadhus and Limits to Religious Mobilisation in North India. *Economic and Political Weekly* 32, no. 49:3149–58.
1999 Sadhus, Sants and Politics: Religious Mobilization and Communalism in North India. Ph.D. diss., University of Washington.

Varady, R. G.
1989 Land Use and Environmental Change in the Gangetic Plain: Nineteenth Century Human Activity in the Banaras Region. In *Culture and*

Power in Banaras, ed. Sandria Freitag, 229–45. Berkeley: University of California Press.

Varun, Dangli Prasad

1981 *Gazetteer of India-Uttar Pradesh District Saharanpur.* Lucknow: Government of Uttar Pradesh, Department of District Gazetteers.

Vatsyayan, Kapila

1992 Ecology and Indian Myth. In *Indigenous Visions: Peoples of India, Attitudes to the Environment*, ed. India International Centre, 1–2. New Delhi: India International Centre.

Vayda, Andrew, and Bradley B. Walters

1999 Against Political Ecology. *Human Ecology* 27, no. 1 (March):167–79.

Vidyarthi, L. P., M. Jha, and B. N. Saraswati

1979 *The Sacred Complex of Kashi.* Delhi: Concept Publishing.

Vishwa Hindu Parisad

1995 Ekatmata Yatra. October/November. Memorandum.

Walsh, Edward, and Rex H. Warland

1983 Social Movement Involvement in the Wake of a Nuclear Accident: Activists and Free Riders in the TMI Area. *American Sociological Review* 48:764–80.

Williams, Brett

2001 A River Runs Through Us. *American Anthropologist* 103, no. 2:409–31.

Wolf, Eric R.

1999 Cognizing "Cognized Models." *American Anthropologist* 101, no. 1:19–22.

Wood, Patricia A.

1982 The Environmentalist Movement: Its Crystallization, Development and Impact. In *Social Movements: Development, Participation, and Dynamics*, ed. James L. Wood and Maurice Jackson, 201–19. Belmont: Wadsworth.

World Health Organization

1984–85 *Guidelines for Drinking Water Quality.* Geneva: World Health Organization. Vols. 1–3.

Zald, Mayer, and John McCarthy, eds.

1979 *The Dynamics of Social Movements.* Cambridge: Winthrop.

Index